Student Solutions Manual

Introduction to Probability and Statistics

FOURTEENTH EDITION

William Mendenhall, III

Robert J, Beaver
University of California, Riverside

Barbara M. Beaver
University of California, Riverside

Prepared by

Barbara M. Beaver
University of California, Riverside

BROOKS/COLE
CENGAGE Learning·

Australia • Brazil • Japan • Korea • Mexico • Singapore • Spain • United Kingdom • United States

For product information and technology assistance, contact us at
**Cengage Learning Customer & Sales Support,
1-800-354-9706**

For permission to use material from this text or product, submit all requests online at **www.cengage.com/permissions**
Further permissions questions can be emailed to
permissionrequest@cengage.com

ISBN-13: 978-1-133-11151-1
ISBN-10: 1-133-11151-3

Brooks/Cole
20 Channel Center Street
Boston, MA 02210
USA

Cengage Learning is a leading provider of customized learning solutions with office locations around the globe, including Singapore, the United Kingdom, Australia, Mexico, Brazil, and Japan. Locate your local office at: **www.cengage.com/global**

Cengage Learning products are represented in Canada by Nelson Education, Ltd.

To learn more about Brooks/Cole, visit **www.cengage.com/brookscole**

Purchase any of our products at your local college store or at our preferred online store **www.cengagebrain.com**

Printed in the United States of America
1 2 3 4 5   17 16 15 14 13

CONTENTS

1: Describing Data with Graphs

1.1 **a** The experimental unit, the individual or object on which a variable is measured, is the student.
b The experimental unit on which the number of errors is measured is the exam.
c The experimental unit is the patient.
d The experimental unit is the azalea plant.
e The experimental unit is the car.

1.3 **a** "Population" is a *discrete* variable because it can take on only integer values.
b "Weight" is a *continuous* variable, taking on any values associated with an interval on the real line.
c "Time" is a *continuous* variable.
d "Number of consumers" is integer-valued and hence *discrete*.

1.5 **a** The experimental unit, the item or object on which variables are measured, is the vehicle.
b Type (qualitative); make (qualitative); carpool or not? (qualitative); one-way commute distance (quantitative continuous); age of vehicle (quantitative continuous)
c Since five variables have been measured, this is *multivariate data*.

1.7 The population of interest consists of voter opinions (for or against the candidate) <u>at the time of the election</u> for all persons voting in the election. Note that when a sample is taken (at some time prior or the election), we are not actually sampling from the population of interest. As time passes, voter opinions change. Hence, the population of voter opinions changes with time, and the sample may not be representative of the population of interest.

1.9 **a** The variable "reading score" is a quantitative variable, which is probably integer-valued and hence discrete.
b The individual on which the variable is measured is the student.
c The population is hypothetical – it does not exist in fact – but consists of the reading scores for all students who could possibly be taught by this method.

1.11 **a-b** The experimental unit is the pair of jeans, on which the qualitative variable "state" is measured.
c-d Construct a statistical table to summarize the data. The pie chart is constructed by partitioning the circle into four parts, according to the total contributed by each part. Since the total number of pairs of jeans is 25, the total number produced in CA represents $9/25 = .36$ or 36% of the total. Thus, this category will be represented by a sector angle of $0.36(360) = 129.6°$. The other sector angles are shown below. The pie chart is shown in the figure that follows.

State	Frequency	Fraction of Total	Sector Angle
CA	9	.36	129.6
AZ	8	.32	115.2
TX	8	.32	115.2

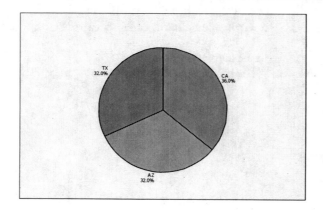

1

The bar chart represents each category as a bar with height equal to the frequency of occurrence of that category and is shown in the figure below.

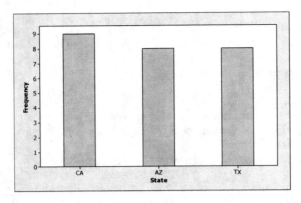

e   From the table or the chart, Texas produced $8/25 = 0.32$ of the jeans.

f   The highest bar represents California, which produced the most pairs of jeans.

g   Since the bars and the sectors are almost equal in size, the three states produced roughly the same number of pairs of jeans.

1.13   **a**   The percentages given in the exercise only add to 94%. We should add another category called "Other", which will account for the other 6% of the responses.

b   Either type of chart is appropriate. Since the data is already presented as percentages of the whole group, we choose to use a pie chart, shown in the figure below.

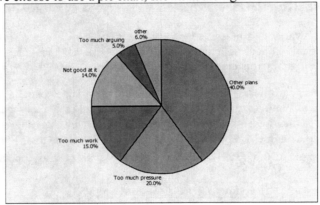

c-d   Answers will vary.

1.15   **a**   The total percentage of responses given in the table is only $(40 + 34 + 19)\% = 93\%$. Hence there are 7% of the opinions not recorded, which should go into a category called "Other" or "More than a few days".

b   Yes. The bars are very close to the correct proportions.

c   Similar to previous exercises. The pie chart is shown on the next page. The bar chart is probably more interesting to look at.

2

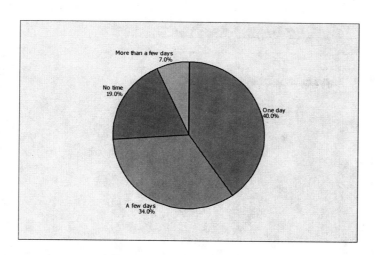

1.17 **a** For $n = 5$, use between 8 and 10 classes.

b

Class i	Class Boundaries	Tally	f_i	Relative frequency, f_i/n
1	1.6 to < 2.1	11	2	.04
2	2.1 to < 2.6	11111	5	.10
3	2.6 to < 3.1	11111	5	.10
4	3.1 to < 3.6	11111	5	.10
5	3.6 to < 4.1	11111 11111 1111	14	.28
6	4.1 to < 4.6	11111 11	7	.14
7	4.6 to < 5.1	11111	5	.10
8	5.1 to < 5.6	11	2	.04
9	5.6 to < 6.1	111	3	.06
10	6.1 to < 6.6	11	2	.04

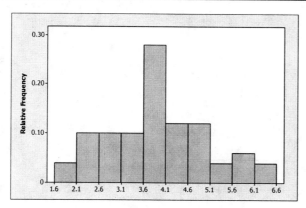

c From **b**, the fraction less than 5.1 is that fraction lying in classes 1-7, or
$$(2 + 5 + \cdots + 7 + 5)/50 = 43/50 = 0.86$$

d From **b**, the fraction larger than 3.6 lies in classes 5-10, or.
$$(14 + 7 + \cdots + 3 + 2)/50 = 33/50 = 0.66$$

e The stem and leaf display has a more peaked mound-shaped distribution than the relative frequency histogram because of the smaller number of groups.

1.19 **a** Since the variable of interest can only take the values 0, 1, or 2, the classes can be chosen as the integer values 0, 1, and 2. The table below shows the classes, their corresponding frequencies and their relative frequencies. The relative frequency histogram follows.

3

Value	Frequency	Relative Frequency
0	5	.25
1	9	.45
2	6	.30

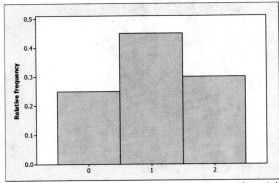

b   Using the table in part **a**, the proportion of measurements greater then 1 is the same as the proportion of "2"s, or 0.30.

c   The proportion of measurements less than 2 is the same as the proportion of "0"s and "1"s, or $0.25 + 0.45 = .70$.

d   The probability of selecting a "2" in a random selection from these twenty measurements is $6/20 = .30$.

e   There are no outliers in this relatively symmetric, mound-shaped distribution.

1.21   The line chart plots "day" on the horizontal axis and "time" on the vertical axis. The line chart shown below reveals that learning is taking place, since the time decreases each successive day.

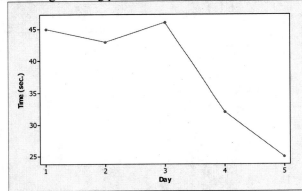

1.23   The dotplot is shown below.

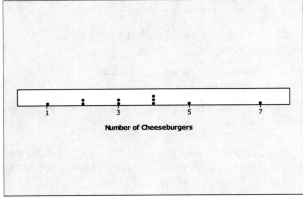

a   The distribution is somewhat mound-shaped (as much as a small set can be); there are no outliers.

    b $2/10 = 0.2$

1.25   **a**   There are a few extremely small numbers, indicating that the distribution is probably skewed to the left.

    b   The range of the data $165 - 8 = 157$. We choose to use seven class intervals of length 25, with subintervals 0 to < 25, 25 to < 50, 50 to < 75, and so on. The tally and relative frequency histogram are shown below.

Class i	Class Boundaries	Tally	f_i	Relative frequency, f_i/n
1	0 to < 25	11	2	2/20
2	25 to < 50		0	0/20
3	50 to < 75	111	3	3/20
4	75 to < 100	111	3	3/20
5	100 to < 125	11	2	2/20
6	125 to < 150	11111 11	7	7/20
7	150 to < 175	111	3	3/20

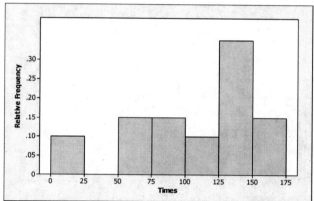

    c   The distribution is indeed skewed left with two possible outliers $- x = 8$ and $x = 11$.

1.27   **a**   The data represent the median weekly earnings for six different levels of education. A bar chart would be the most appropriate graphical method.

    b   The bar chart is shown below.

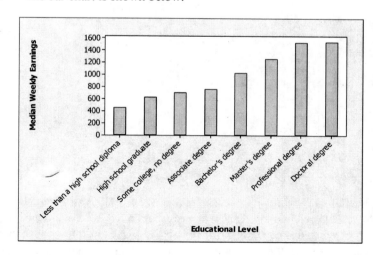

    c   The median weekly earnings increases substantially as the person's educational level increases.

5

1.29 **a** Similar to previous exercises. The pie chart is shown below.

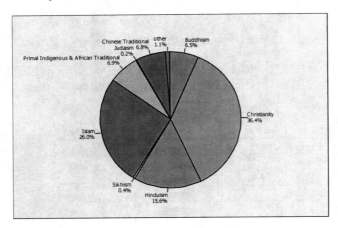

b The bar chart is shown below.

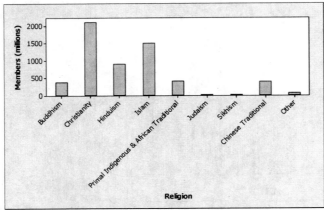

c The Pareto chart is a bar chart with the heights of the bars ordered from large to small. This display is more effective than the pie chart.

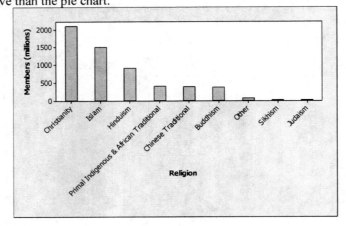

1.31 **a** The data ranges from .2 to 5.2, or 5.0 units. Since the number of class intervals should be between five and twelve, we choose to use eleven class intervals, with each class interval having length 0.50 ($5.0/11 = .45$, which, rounded to the nearest convenient fraction, is .50). We must now select interval boundaries such that no measurement can fall on a boundary point. The subintervals .1 to < .6, .6 to < 1.1, and so on, are convenient and a tally is constructed.

6

Class i	Class Boundaries	Tally	f_i	Relative frequency, f_i/n
1	0.1 to < 0.6	11111 11111	10	.167
2	0.6 to < 1.1	11111 11111 11111	15	.250
3	1.1 to < 1.6	11111 11111 11111	15	.250
4	1.6 to < 2.1	11111 11111	10	.167
5	2.1 to < 2.6	1111	4	.067
6	2.6 to < 3.1	1	1	.017
7	3.1 to < 3.6	11	2	.033
8	3.6 to < 4.1	1	1	.017
9	4.1 to < 4.6	1	1	.017
10	4.6 to < 5.1		0	.000
11	5.1 to < 5.6	1	1	.017

The relative frequency histogram is shown below.

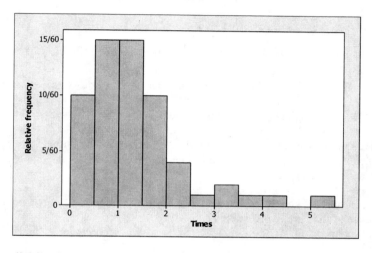

The distribution is skewed to the right, with several unusually large observations.

b For some reason, one person had to wait 5.2 minutes. Perhaps the supermarket was understaffed that day, or there may have been an unusually large number of customers in the store.

c The two graphs convey the same information. The stem and leaf plot allows us to actually recreate the actual data set, while the histogram does not.

1.33 **a** Answers will vary.

b The stem and leaf plot is constructed using the tens place as the stem and the ones place as the leaf. *Minitab* divides each stem into two parts to create a better descriptive picture. Notice that the distribution is roughly mound-shaped.

Stem-and-Leaf Display: Ages
```
Stem-and-leaf of Ages   N  = 38
Leaf Unit = 1.0

 2    4   69
 3    5   3
 7    5   6678
13    6   003344
19    6   567778
19    7   011234
13    7   7889
 9    8   013
 6    8   58
 4    9   0033
```

c Three of the five youngest presidents – Kennedy, Lincoln and Garfield – were assassinated while in office. This would explain the fact that their ages at death were in the lower tail of the distribution.

7

1.35   **a**   Histograms will vary from student to student. A typical histogram, generated by *Minitab* is shown below.

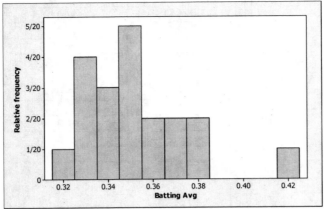

b   Since 1 of the 20 players has an average above 0.400, the chance is 1 out of 20 or $1/20 = 0.05$.

1.37   **a**   The variable being measured is a discrete variable – the number of hazardous waste sites in each of the 50 United States.
b   The distribution is skewed to the right, with a several unusually large measurements. The five states marked as HI are California, Michigan, New Jersey, New York and Pennsylvania.
c   Four of the five states are quite large in area, which might explain the large number of hazardous waste sites. However, the fifth state is relatively small, and other large states do not have unusually large number of waste sites. The pattern is not clear.

1.39   To determine whether a distribution is likely to be skewed, look for the likelihood of observing extremely large or extremely small values of the variable of interest.
a   The distribution of non-secured loan sizes might be skewed (a few extremely large loans are possible).
b   The distribution of secured loan sizes is not likely to contain unusually large or small values.
c   Not likely to be skewed.
d   Not likely to be skewed.
e   If a package is dropped, it is likely that all the shells will be broken. Hence, a few large number of broken shells is possible. The distribution will be skewed.
f   If an animal has one tick, he is likely to have more than one. There will be some "0"s with uninfected rabbits, and then a larger number of large values. The distribution will not be symmetric.

1.41   **a**   Weight is continuous, taking any positive real value.
b   Body temperature is continuous, taking any real value.
c   Number of people is discrete, taking the values 0, 1, 2, …
d   Number of properties is discrete.
e   Number of claims is discrete.

1.43   Stem and leaf displays may vary from student to student. The most obvious choice is to use the tens digit as the stem and the ones digit as the leaf.

```
 7 | 8 9
 8 | 0 1 7
 9 | 0 1 2 4 4 5 6 6 6 8 8
10 | 1 7 9
11 | 2
```
The display is fairly mound-shaped, with a large peak in the middle.

8

1.45   a-b   Answers will vary from student to student. The students should notice that the distribution is skewed to the right with a few pennies being unusually old. A typical histogram is shown below.

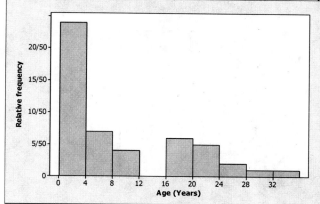

1.47   Answers will vary from student to student. The students should notice that the distribution is skewed to the right with a few presidents (Truman, Cleveland, and F.D. Roosevelt) casting an unusually large number of vetoes.

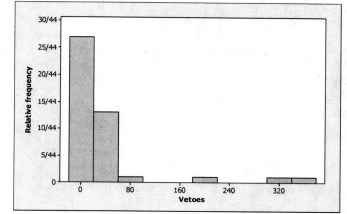

1.49   a   The line chart is shown below. The year in which a horse raced does not appear to have an effect on his winning time.

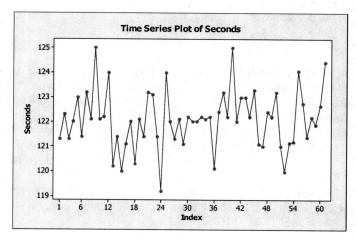

b   Since the year of the race is not important in describing the data set, the distribution can be described using a relative frequency histogram. The distribution that follows is roughly mound-shaped with an unusually fast ($x = 119.2$) race times the year that *Secretariat* won the derby.

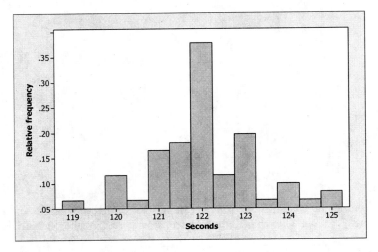

1.51   **a**   The popular vote within each state should vary depending on the size of the state. Since there are several very large states (in population) in the United States, the distribution should be skewed to the right. **b-c** Histograms will vary from student to student, but should resemble the histogram generated by *Minitab* in the figure that follows. The distribution is indeed skewed to the right, with one "outlier" – California (and possibly Florida and New York).

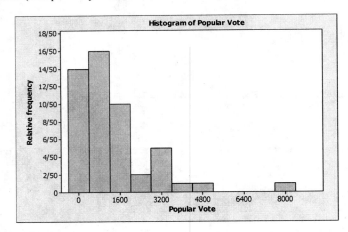

1.53   **a-b** Popular vote is skewed to the right while the percentage of popular vote is roughly mound-shaped. While the distribution of popular vote has outliers (California, Florida and New York), there are no outliers in the distribution of percentage of popular vote. When the stem and leaf plots are turned 90°, the shapes are very similar to the histograms.
c   Once the size of the state is removed by calculating the percentage of the popular vote, the unusually large values in the set of "popular votes" will disappear, and each state will be measured on an equal basis. The data then distribute themselves in a mound-shape around the average percentage of the popular vote.

1.55   **a-b** Answers will vary from student to student. Since the graph gives a range of values for Zimbabwe's share, we have chosen to use the 13% figure, and have used 3% in the "Other" category. The pie chart and bar charts are shown on the next page.

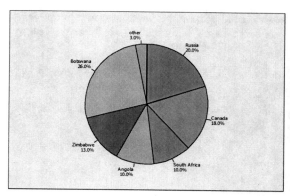
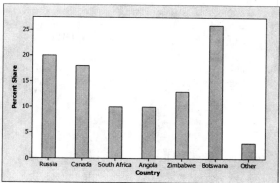

c-d The Pareto chart is shown below. Either the pie chart or the Pareto chart is more effective than the bar chart.

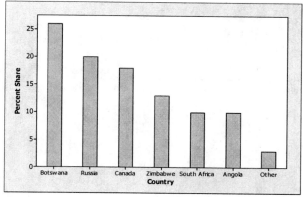

1.57   The relative frequency histogram below was constructed using classes of length 1.0 starting at $x = 0.0$.

Class i	Class Boundaries	Tally	f_i	Relative frequency, f_i/n
1	0.0 to < 1.0	111	3	3/41
2	1.0 to < 2.0	1	1	1/41
3	2.0 to < 3.0	1	1	1/41
4	3.0 to < 4.0	111	3	3/41
5	4.0 to < 5.0	111	3	3/41
6	5.0 to < 6.0	1111	4	4/41
7	6.0 to < 7.0	1111	4	4/41
8	7.0 to < 8.0	11111 1	6	6/41
9	8.0 to < 9.0	11111 1	6	6/41
10	9.0 to < 10.0	11111 11111	10	10/41

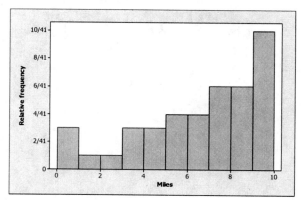

11

a   The distribution is skewed to the left, with an unusual peak in the first class (within one mile of UCR).

b   As the distance from UCR increases, each successive area increases in size, thus allowing for more Starbucks stores in that region.

1.59   **a-b** Answers will vary. A typical histogram is shown below. Notice the gaps and the bimodal nature of the histogram, probably due to the fact that the samples were collected at different locations.

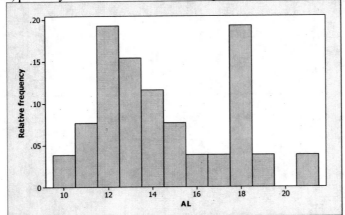

c   The dotplot is shown below. The locations are indeed responsible for the unusual gaps and peaks in the relative frequency histogram.

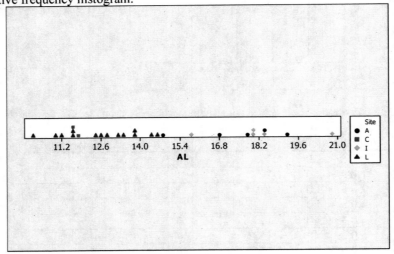

1.61   Answers will vary from student to student. Students should notice that both distributions are skewed left. The higher peak with a low bar to its left in the laptop group may indicate that students who would generally receive average scores (65-75) are scoring higher than usual. This may or may not be *caused* by the fact that they used laptop computers.

1.63   **a-b** The *Minitab* stem and leaf plot is shown below. The distribution is slightly skewed to the right.

Stem-and-Leaf Display: Tax

```
Stem-and-leaf of Tax   N  = 51
Leaf Unit = 1.0

    1    2  6
    3    3  22
   16    3  5557778888999
 (15)    4  000011111223333
   20    4  566689
   14    5  00111234
    6    5  58
    4    6  133
    1    6  7
```

c   Arkansas (26.4), Wyoming (32.4) and New Jersey (32.9) have gasoline taxes that are somewhat smaller than most, but they are not "outliers" in the sense that they lie far away from the rest of the measurements in the data set.

1.65   The data should be displayed with either a bar chart or a pie chart. The pie chart is shown below.

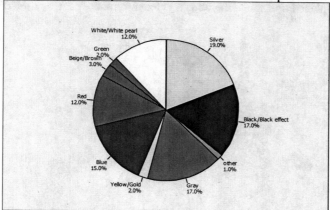

1.67   **a-b** The distribution is approximately mound-shaped, with one unusual measurement, in the class with midpoint at 100.8°. Perhaps the person whose temperature was 100.8 has some sort of illness coming on?

        c   The value 98.6° is slightly to the right of center.

13

2: Describing Data with Numerical Measures

2.1   **a**   The dotplot shown below plots the five measurements along the horizontal axis. Since there are two "1"s, the corresponding dots are placed one above the other. The approximate center of the data appears to be around 1.

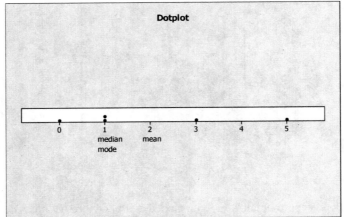

     b   The mean is the sum of the measurements divided by the number of measurements, or

$$\bar{x} = \frac{\sum x_i}{n} = \frac{0+5+1+1+3}{5} = \frac{10}{5} = 2$$

To calculate the median, the observations are first ranked from smallest to largest: 0, 1, 1, 3, 5. Then since $n = 5$, the position of the median is $0.5(n+1) = 3$, and the median is the 3rd ranked measurement, or $m = 1$. The mode is the measurement occurring most frequently, or mode = 1.

     c   The three measures in part **b** are located on the dotplot. Since the median and mode are to the left of the mean, we conclude that the measurements are skewed to the right.

2.3   **a** $\bar{x} = \dfrac{\sum x_i}{n} = \dfrac{58}{10} = 5.8$

     b   The ranked observations are: 2, 3, 4, 5, 5, 6, 6, 8, 9, 10. Since $n = 10$, the median is halfway between the 5th and 6th ordered observations, or $m = (5+6)/2 = 5.5$.

     c   There are two measurements, 5 and 6, which both occur twice. Since this is the highest frequency of occurrence for the data set, we say that the set is *bimodal* with modes at 5 and 6.

2.5   **a**   Although there may be a few households who own more than one DVR, the majority should own either 0 or 1. The distribution should be slightly skewed to the right.

     b   Since most households will have only one DVR, we guess that the mode is 1.

     c   The mean is

$$\bar{x} = \frac{\sum x_i}{n} = \frac{1+0+\cdots+1}{25} = \frac{27}{25} = 1.08$$

To calculate the median, the observations are first ranked from smallest to largest: There are six 0s, thirteen 1s, four 2s, and two 3s. Then since $n = 25$, the position of the median is $0.5(n+1) = 13$, which is the 13th ranked measurement, or $m = 1$. The mode is the measurement occurring most frequently, or mode = 1.

     d   The relative frequency histogram is shown on the next page, with the three measures superimposed. Notice that the mean falls slightly to the right of the median and mode, indicating that the measurements are slightly skewed to the right.

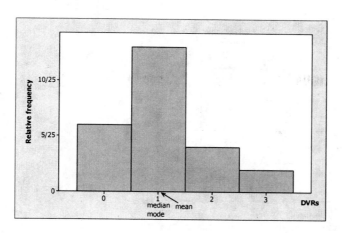

2.7   It is obvious that any one family cannot have 2.5 children, since the number of children per family is a quantitative discrete variable. The researcher is referring to the average number of children per family calculated for all families in the United States during the 1930s. The average does not necessarily have to be integer-valued.

2.9   The distribution of sports salaries will be skewed to the right, because of the very high salaries of some sports figures. Hence, the median salary would be a better measure of center than the mean.

2.11   **a**   Similar to previous exercises.

$$\bar{x} = \frac{\sum x_i}{n} = \frac{115}{21} = 5.476$$

The ranked observations are:

1	2	3	6	10	18
1	2	3	6	10	
1	2	4	7	11	
1	2	5	8	12	

The position of the median is $0.5(n+1) = 11$, and the median is the 11th observation or $m = 4$. There are two observations, 1 and 2, both of which occur four times. Hence, the data set is *bimodal*—it has two modes, 1 and 2.

b   Since the mean is larger than the median, the data are skewed to the right.

c   The dotplot is shown below. The distribution is skewed to the right.

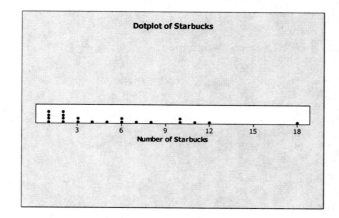

2.13   **a** $\bar{x} = \frac{\sum x_i}{n} = \frac{12}{5} = 2.4$

b Create a table of differences, $\left(x_i - \bar{x}\right)$ and their squares, $\left(x_i - \bar{x}\right)^2$.

x_i	$x_i - \bar{x}$	$\left(x_i - \bar{x}\right)^2$
2	–0.4	0.16
1	–1.4	1.96
1	–1.4	1.96
3	0.6	0.36
5	2.6	6.76
Total	0	11.20

Then

$$s^2 = \frac{\sum\left(x_i - \bar{x}\right)^2}{n-1} = \frac{(2-2.4)^2 + \cdots + (5-2.4)^2}{4} = \frac{11.20}{4} = 2.8$$

c The sample standard deviation is the positive square root of the variance or

$$s = \sqrt{s^2} = \sqrt{2.8} = 1.673$$

d Calculate $\sum x_i^2 = 2^2 + 1^2 + \cdots + 5^2 = 40$. Then

$$s^2 = \frac{\sum x_i^2 - \dfrac{\left(\sum x_i\right)^2}{n}}{n-1} = \frac{40 - \dfrac{(12)^2}{5}}{4} = \frac{11.2}{4} = 2.8 \text{ and } s = \sqrt{s^2} = \sqrt{2.8} = 1.673.$$

The results of parts **a** and **b** are identical.

2.15 **a** The range is $R = 4 - 1 = 3$.   **b** $\bar{x} = \dfrac{\sum x_i}{n} = \dfrac{17}{8} = 2.125$

c Calculate $\sum x_i^2 = 4^2 + 1^2 + \cdots + 2^2 = 45$. Then

$$s^2 = \frac{\sum x_i^2 - \dfrac{\left(\sum x_i\right)^2}{n}}{n-1} = \frac{45 - \dfrac{(17)^2}{8}}{7} = \frac{8.875}{7} = 1.2679 \text{ and } s = \sqrt{s^2} = \sqrt{1.2679} = 1.126.$$

2.17 **a** The range is $R = 2.39 - 1.28 = 1.11$.

b Calculate $\sum x_i^2 = 1.28^2 + 2.39^2 + \cdots + 1.51^2 = 15.415$. Then

$$s^2 = \frac{\sum x_i^2 - \dfrac{\left(\sum x_i\right)^2}{n}}{n-1} = \frac{15.415 - \dfrac{(8.56)^2}{5}}{4} = \frac{.76028}{4} = .19007$$

and $s = \sqrt{s^2} = \sqrt{.19007} = .436$

c The range, $R = 1.11$, is $1.11/.436 = 2.5$ standard deviations.

2.19 **a** The range of the data is $R = 6 - 1 = 5$ and the range approximation with $n = 10$ is

$$s \approx \frac{R}{3} = 1.67$$

b The standard deviation of the sample is

$$s = \sqrt{s^2} = \sqrt{\frac{\sum x_i^2 - \dfrac{\left(\sum x_i\right)^2}{n}}{n-1}} = \sqrt{\frac{130 - \dfrac{(32)^2}{10}}{9}} = \sqrt{3.0667} = 1.751$$

which is very close to the estimate for part **a**.

c-e From the dotplot on the next page, you can see that the data set is not mound-shaped. Hence you can use Tchebysheff's Theorem, but not the Empirical Rule to describe the data.

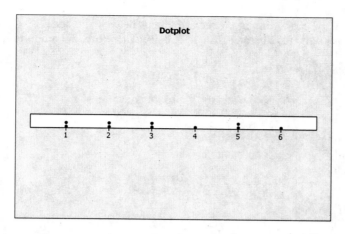

2.21   **a**   The interval from 40 to 60 represents $\mu \pm \sigma = 50 \pm 10$. Since the distribution is relatively mound-shaped, the proportion of measurements between 40 and 60 is .68 (or 68%) according to the Empirical Rule and is shown below.

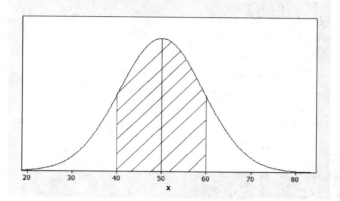

b   Again, using the Empirical Rule, the interval $\mu \pm 2\sigma = 50 \pm 2(10)$ or between 30 and 70 contains approximately .95 (or 95%) of the measurements.

c   Refer to the figure below.

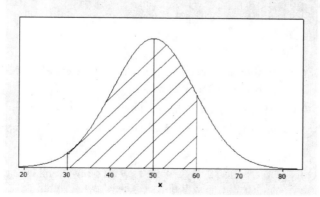

Since approximately 68% of the measurements are between 40 and 60, the symmetry of the distribution implies that 34% of the measurements are between 50 and 60. Similarly, since 95% of the measurements are between 30 and 70, approximately 47.5% are between 30 and 50. Thus, the proportion of measurements between 30 and 60 is

$$0.34 + 0.475 = 0.815$$

17

d   From the figure in part **a**, the proportion of the measurements between 50 and 60 is 0.34 and the proportion of measurements which are greater than 50 is 0.50. Therefore, the proportion that is greater than 60 must be

$$0.5 - 0.34 = 0.16$$

2.23   **a**   The range of the data is $R = 1.1 - 0.5 = 0.6$ and the approximate value of s is

$$s \approx \frac{R}{3} = 0.2$$

b   Calculate $\sum x_i = 7.6$ and $\sum x_i^2 = 6.02$, the sample mean is

$$\bar{x} = \frac{\sum x_i}{n} = \frac{7.6}{10} = .76$$

and the standard deviation of the sample is

$$s = \sqrt{s^2} = \sqrt{\frac{\sum x_i^2 - \frac{\left(\sum x_i\right)^2}{n}}{n-1}} = \sqrt{\frac{6.02 - \frac{(7.6)^2}{10}}{9}} = \sqrt{\frac{0.244}{9}} = 0.165$$

which is very close to the estimate from part **a**.

2.25   According to the Empirical Rule, if a distribution of measurements is approximately mound-shaped,
   a   approximately 68% or 0.68 of the measurements fall in the interval $\mu \pm \sigma = 12 \pm 2.3$ or 9.7 to 14.3
   b   approximately 95% or 0.95 of the measurements fall in the interval $\mu \pm 2\sigma = 12 \pm 4.6$ or 7.4 to 16.6
   c   approximately 99.7% or 0.997 of the measurements fall in the interval $\mu \pm 3\sigma = 12 \pm 6.9$ or 5.1 to 18.9
   Therefore, approximately 0.3% or 0.003 will fall outside this interval.

2.27   **a**   The relative frequency histogram is shown below.

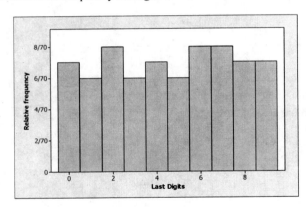

The distribution is relatively "flat" or "uniformly distributed" between 0 and 9. Hence, the center of the distribution should be approximately halfway between 0 and 9 or $(0+9)/2 = 4.5$.
   b   The range of the data is $R = 9 - 0 = 9$. Using the range approximation, $s \approx R/4 = 9/4 = 2.25$.
   c   Using the data entry method the students should find $\bar{x} = 4.586$ and $s = 2.892$, which are fairly close to our approximations.

2.29   **a**   Although most of the animals will die at around 32 days, there may be a few animals that survive a very long time, even with the infection. The distribution will probably be skewed right.

   b   Using Tchebysheff's Theorem, at least 3/4 of the measurements should be in the interval $\mu \pm \sigma \Rightarrow 32 \pm 72$ or 0 to 104 days.

2.31 **a** We choose to use 12 classes of length 1.0. The tally and the relative frequency histogram follow.

Class i	Class Boundaries	Tally	f_i	Relative frequency, f_i/n
1	2 to < 3	1	1	1/70
2	3 to < 4	1	1	1/70
3	4 to < 5	111	3	3/70
4	5 to < 6	11111	5	5/70
5	6 to < 7	11111	5	5/70
6	7 to < 8	11111 11111 11	12	12/70
7	8 to < 9	11111 11111 11111 111	18	18/70
8	9 to < 10	11111 11111 11111	15	15/70
9	10 to < 11	11111 1	6	6/70
10	11 to < 12	111	3	3/70
11	12 to < 13		0	0
12	13 to < 14	1	1	1/70

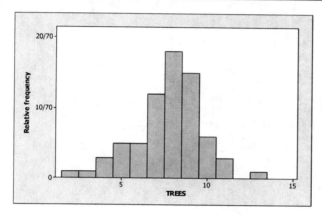

b Calculate $n = 70$, $\sum x_i = 541$ and $\sum x_i^2 = 4453$. Then $\bar{x} = \dfrac{\sum x_i}{n} = \dfrac{541}{70} = 7.729$ is an estimate of μ.

c The sample standard deviation is

$$s = \sqrt{\frac{\sum x_i^2 - \dfrac{\left(\sum x_i\right)^2}{n}}{n-1}} = \sqrt{\frac{4453 - \dfrac{(541)^2}{70}}{69}} = \sqrt{3.9398} = 1.985$$

The three intervals, $\bar{x} \pm ks$ for $k = 1, 2, 3$ are calculated below. The table shows the actual percentage of measurements falling in a particular interval as well as the percentage predicted by Tchebysheff's Theorem and the Empirical Rule. Note that the Empirical Rule should be fairly accurate, as indicated by the mound-shape of the histogram in part **a**.

k	$\bar{x} \pm ks$	Interval	Fraction in Interval	Tchebysheff	Empirical Rule
1	7.729 ± 1.985	5.744 to 9.714	50/70 = 0.71	at least 0	≈ 0.68
2	7.729 ± 3.970	3.759 to 11.699	67/70 = 0.96	at least 0.75	≈ 0.95
3	7.729 ± 5.955	1.774 to 13.684	70/70 = 1.00	at least 0.89	≈ 0.997

2.33 **a-b** Calculate $R = 93 - 51 = 42$ so that $s \approx R/4 = 42/4 = 10.5$.

c Calculate $n = 30$, $\sum x_i = 2145$ and $\sum x_i^2 = 158,345$. Then

$$s^2 = \frac{\sum x_i^2 - \dfrac{\left(\sum x_i\right)^2}{n}}{n-1} = \frac{158,345 - \dfrac{(2145)^2}{30}}{29} = 171.6379 \text{ and } s = \sqrt{171.6379} = 13.101$$

which is fairly close to the approximate value of s from part **b**.

19

d The two intervals are calculated below. The proportions agree with Tchebysheff's Theorem, but are not to close to the percentages given by the Empirical Rule. (This is because the distribution is not quite mound-shaped.)

k	$\bar{x} \pm ks$	Interval	Fraction in Interval	Tchebysheff	Empirical Rule
2	71.5 ± 26.20	45.3 to 97.7	$30/30 = 1.00$	at least 0.75	≈ 0.95
3	71.5 ± 39.30	32.2 to 110.80	$30/30 = 1.00$	at least 0.89	≈ 0.997

2.35 **a** Calculate $R = 2.39 - 1.28 = 1.11$ so that $s \approx R/2.5 = 1.11/2.5 = .444$.

b In Exercise 2.17, we calculated $\sum x_i = 8.56$ and $\sum x_i^2 = 1.28^2 + 2.39^2 + \cdots + 1.51^2 = 15.415$. Then

$$s^2 = \frac{\sum x_i^2 - \frac{(\sum x_i)^2}{n}}{n-1} = \frac{15.451 - \frac{(8.56)^2}{5}}{4} = \frac{.76028}{4} = .19007$$

and $s = \sqrt{s^2} = \sqrt{.19007} = .436$, which is very close to our estimate in part **a**.

2.37 **a** Calculate $n = 15$, $\sum x_i = 21$ and $\sum x_i^2 = 49$. Then $\bar{x} = \frac{\sum x_i}{n} = \frac{21}{15} = 1.4$ and

$$s^2 = \frac{\sum x_i^2 - \frac{(\sum x_i)^2}{n}}{n-1} = \frac{49 - \frac{(21)^2}{15}}{14} = 1.4$$

b Using the frequency table and the grouped formulas, calculate

$$\sum x_i f_i = 0(4) + 1(5) + 2(2) + 3(4) = 21$$
$$\sum x_i^2 f_i = 0^2(4) + 1^2(5) + 2^2(2) + 3^2(4) = 49$$

Then, as in part **a**,

$$\bar{x} = \frac{\sum x_i f_i}{n} = \frac{21}{15} = 1.4$$

$$s^2 = \frac{\sum x_i^2 f_i - \frac{(\sum x_i f_i)^2}{n}}{n-1} = \frac{49 - \frac{(21)^2}{15}}{14} = 1.4$$

2.39 **a** The data in this exercise have been arranged in a frequency table.

x_i	0	1	2	3	4	5	6	7	8	9	10
f_i	10	5	3	2	1	1	1	0	0	1	1

Using the frequency table and the grouped formulas given in Exercise 2.37, calculate

$$\sum x_i f_i = 0(10) + 1(5) + \cdots + 10(1) = 51$$
$$\sum x_i^2 f_i = 0^2(10) + 1^2(5) + \cdots + 10^2(1) = 293$$

Then

$$\bar{x} = \frac{\sum x_i f_i}{n} = \frac{51}{25} = 2.04$$

$$s^2 = \frac{\sum x_i^2 f_i - \frac{(\sum x_i f_i)^2}{n}}{n-1} = \frac{293 - \frac{(51)^2}{25}}{24} = 7.873 \text{ and } s = \sqrt{7.873} = 2.806.$$

b-c The three intervals $\bar{x} \pm ks$ for $k = 1, 2, 3$ are calculated in the table along with the actual proportion of measurements falling in the intervals. Tchebysheff's Theorem is satisfied and the approximation given by the Empirical Rule are fairly close for $k = 2$ and $k = 3$.

k	$\bar{x} \pm ks$	Interval	Fraction in Interval	Tchebysheff	Empirical Rule
1	2.04 ± 2.806	-0.766 to 4.846	$21/25 = 0.84$	at least 0	≈ 0.68
2	2.04 ± 5.612	-3.572 to 7.652	$23/25 = 0.92$	at least 0.75	≈ 0.95
3	2.04 ± 8.418	-6.378 to 10.458	$25/25 = 1.00$	at least 0.89	≈ 0.997

2.40 The ordered data are:

$$0, 1, 3, 4, 4, 5, 6, 6, 7, 7, 8$$

a With $n = 12$, the median is in position $0.5(n+1) = 6.5$, or halfway between the 6th and 7th observations. The lower quartile is in position $0.25(n+1) = 3.25$ (one-fourth of the way between the 3rd and 4th observations) and the upper quartile is in position $0.75(n+1) = 9.75$ (three-fourths of the way between the 9th and 10th observations). Hence, $m = (5+6)/2 = 5.5$, $Q_1 = 3 + 0.25(4-3) = 3.25$ and $Q_3 = 6 + 0.75(7-6) = 6.75$. Then the five-number summary is

Min	Q_1	Median	Q_3	Max
0	3.25	5.5	6.75	8

and

$$IQR = Q_3 - Q_1 = 6.75 - 3.25 = 3.50$$

b Calculate $n = 12$, $\sum x_i = 57$ and $\sum x_i^2 = 337$. Then $\bar{x} = \dfrac{\sum x_i}{n} = \dfrac{57}{12} = 4.75$ and the sample standard deviation is

$$s = \sqrt{\frac{\sum x_i^2 - \dfrac{(\sum x_i)^2}{n}}{n-1}} = \sqrt{\frac{337 - \dfrac{(57)^2}{12}}{11}} = \sqrt{6.022727} = 2.454$$

c For the smaller observation, $x = 0$,

$$z\text{-score} = \frac{x - \bar{x}}{s} = \frac{0 - 4.75}{2.454} = -1.94$$

and for the largest observation, $x = 8$,

$$z\text{-score} = \frac{x - \bar{x}}{s} = \frac{8 - 4.75}{2.454} = 1.32$$

Since neither z-score exceeds 2 in absolute value, none of the observations are unusually small or large.

2.41 The ordered data are:

$$0, 1, 5, 6, 7, 8, 9, 10, 12, 12, 13, 14, 16, 19, 19$$

With $n = 15$, the median is in position $0.5(n+1) = 8$ (the 8th ordered observation), so that $m = 10$. The lower quartile is in position $0.25(n+1) = 4$ (the 4th ordered observation) so that $Q_1 = 6$ and the upper quartile is in position $0.75(n+1) = 12$ (the 12th ordered observation) so that $Q_3 = 14$. Then the five-number summary is

Min	Q_1	Median	Q_3	Max
0	6	10	14	19

and $IQR = Q_3 - Q_1 = 14 - 6 = 8$.

2.43 Notice that the data is already ranked from smallest to largest.

a-b With $n = 8$, the lower quartile is in position $0.25(n+1) = 2.25$ (one-fourth of the way between the 2nd and 3rd observations) and the upper quartile is in position $0.75(n+1) = 6.75$ (three-fourths of the way between the 6th and 7th observations). Hence, $Q_1 = .30 + 0.25(.35 - .30) = .3125$ and $Q_3 = .58 + 0.75(.76 - .58) = .7150$. Then $IQR = Q_3 - Q_1 = .7150 - .3125 = .4025$.

c The upper and lower fences are then calculated as:

Lower fence = $Q_1 - 1.5IQR = .3125 - 1.5(.4025) = -.29125$

Upper fence = $Q_3 + 1.5IQR = .7150 + 1.5(.4025) = 1.31875$

There are no data points that lie outside these fences.

2.45 The ordered data are:

$$2, 3, 4, 5, 6, 6, 6, 7, 8, 9, 9, 10, 22$$

For $n = 13$, the position of the median is $0.5(n+1) = 0.5(13+1) = 7$ and $m = 6$. The positions of the quartiles are $0.25(n+1) = 3.5$ and $0.75(n+1) = 10.5$, so that $Q_1 = 4.5$, $Q_3 = 9$, and $IQR = 9 - 4.5 = 4.5$. The *lower and upper fences* are:

$$Q_1 - 1.5IQR = 4.5 - 6.75 = -2.25$$
$$Q_3 + 1.5IQR = 9 + 6.75 = 15.75$$

The value $x = 22$ lies outside the upper fence and is an outlier. The box plot is shown below. The lower whisker connects the box to the smallest value that is not an outlier, which happens to be the minimum value, $x = 2$. The upper whisker connects the box to the largest value that is not an outlier or $x = 10$.

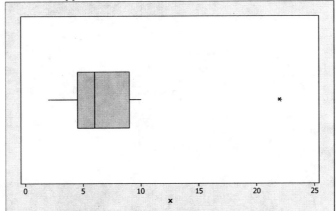

2.47 **a** The ordered data are shown below:

1.70	101.00	209.00	264.00	316.00	445.00
1.72	118.00	218.00	278.00	318.00	481.00
5.90	168.00	221.00	286.00	329.00	485.00
8.80	180.00	241.00	314.00	397.00	
85.40	183.00	252.00	315.00	406.00	

For $n = 28$, the position of the median is $0.5(n+1) = 14.5$ and the positions of the quartiles are $0.25(n+1) = 7.25$ and $0.75(n+1) = 21.75$. The lower quartile is ¼ the way between the 7th and 8th measurements or $Q_1 = 118 + 0.25(168 - 118) = 130.5$ and the upper quartile is ¾ the way between the 21st and 22nd measurements or $Q_3 = 316 + 0.75(318 - 316) = 317.5$. Then the five-number summary is

Min	Q_1	Median	Q_3	Max
1.70	130.5	246.5	317.5	485

b Calculate $IQR = Q_3 - Q_1 = 317.5 - 130.5 = 187$. Then the *lower and upper fences* are:

$$Q_1 - 1.5IQR = 130.5 - 280.5 = -150$$
$$Q_3 + 1.5IQR = 317.5 + 280.5 = 598$$

22

The box plot is shown below. Since there are no outliers, the whiskers connect the box to the minimum and maximum values in the ordered set.

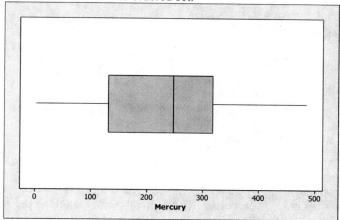

c-d The boxplot does not identify any of the measurements as outliers, mainly because the large variation in the measurements cause the IQR to be large. However, the student should notice the extreme difference in the magnitude of the first four observations taken on young dolphins. These animals have not been alive long enough to accumulate a large amount of mercury in their bodies.

2.49   **a**   For $n = 15$, the position of the median is $0.5(n+1) = 8$ and the positions of the quartiles are $0.25(n+1) = 4$ and $0.75(n+1) = 12$, while for $n = 16$, the position of the median is $0.5(n+1) = 8.5$ and the positions of the quartiles are $0.25(n+1) = 4.25$ and $0.75(n+1) = 12.75$. The sorted measurements are shown below.

        Aaron Rodgers: 7, 12, 15, 18, 19, 19, 19, 21, 21, 22, 25, 26, 27, 27, 34
        Drew Brees: 21, 22, 23, 24, 24, 25, 27, 27, 28, 29, 29, 30, 33, 34, 35, 37

For Aaron Rodgers,
$$m = 21, Q_1 = 18 \text{ and } Q_3 = 26.$$
For Drew Brees,
$$m = (27+28)/2 = 27.5, Q_1 = 24 + 0.25(24 - 24) = 24 \text{ and } Q_3 = 30 + 0.75(33 - 30) = 32.25.$$
Then the five-number summaries are

	Min	Q_1	Median	Q_3	Max
Rodgers	7	18	21	26	34
Brees	21	24	27.5	32.25	37

b   For Aaron Rodgers, calculate $IQR = Q_3 - Q_1 = 26 - 18 = 8$. Then the *lower and upper fences* are:
$$Q_1 - 1.5IQR = 18 - 12 = 6$$
$$Q_3 + 1.5IQR = 26 + 12 = 38$$
and there are no outliers.
For Drew Brees, calculate $IQR = Q_3 - Q_1 = 32.25 - 24 = 8.25$. Then the *lower and upper fences* are:
$$Q_1 - 1.5IQR = 24 - 12.375 = 11.625$$
$$Q_3 + 1.5IQR = 32.25 + 12.375 = 44.625$$
and there are no outliers. The box plots are shown on the next page.

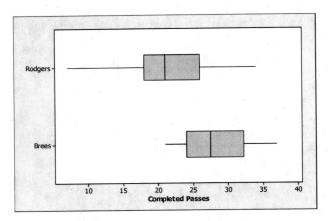

c   Answers will vary. Both distributions are relatively symmetric and somewhat mound-shaped. The Rodgers distribution is slightly more variable; Brees has a higher median number of completed passes.

2.51   **a**   Just by scanning through the 20 measurements, it seems that there are a few unusually small measurements, which would indicate a distribution that is skewed to the left.

    b   The position of the median is $0.5(n+1) = 0.5(25+1) = 10.5$ and $m = (120+127)/2 = 123.5$. The mean is

$$\bar{x} = \frac{\sum x_i}{n} = \frac{2163}{20} = 108.15$$

which is smaller than the median, indicate a distribution skewed to the left.

    c   The positions of the quartiles are $0.25(n+1) = 5.25$ and $0.75(n+1) = 15.75$, so that $Q_1 = 65 - .25(87 - 65) = 70.5$, $Q_3 = 144 + .75(147 - 144) = 146.25$, and $IQR = 146.25 - 70.5 = 75.75$. The *lower and upper fences* are:

$$Q_1 - 1.5IQR = 70.5 - 113.625 = -43.125$$
$$Q_3 + 1.5IQR = 146.25 + 113.625 = 259.875$$

The box plot is shown below. There are no outliers. The long left whisker and the median line located to the right of the center of the box indicates that the distribution that is skewed to the left.

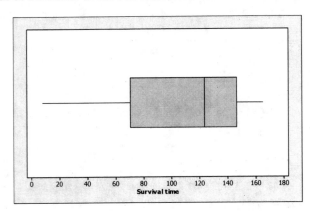

2.53   Answers will vary. The student should notice the outliers in the female group, and that the median female temperature is higher than the median male temperature.

2.55   **a**   The ordered sets are shown below:

Generic					Sunmaid				
24	25	25	25	26	22	24	24	24	24
26	26	26	26	27	25	25	27	28	28
27	28	28	28		28	28	29	30	

24

For $n = 14$, the position of the median is $0.5(n+1) = 0.5(14+1) = 7.5$ and the positions of the quartiles are $0.25(n+1) = 3.75$ and $0.75(n+1) = 11.25$, so that

Generic: $m = 26$, $Q_1 = 25$, $Q_3 = 27.25$, and $IQR = 27.25 - 25 = 2.25$

Sunmaid: $m = 26$, $Q_1 = 24$, $Q_3 = 28$, and $IQR = 28 - 24 = 4$

b **Generic:** *Lower and upper fences* are:
$$Q_1 - 1.5IQR = 25 - 3.375 = 21.625$$
$$Q_3 + 1.5IQR = 27.25 + 3.375 = 30.625$$

Sunmaid: *Lower and upper fences* are:
$$Q_1 - 1.5IQR = 24 - 6 = 18$$
$$Q_3 + 1.5IQR = 28 + 6 = 34$$

The box plots are shown below. There are no outliers.

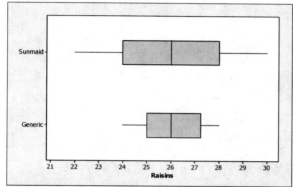

d If the boxes are not being underfilled, the average size of the raisins is roughly the same for the two brands. However, since the number of raisins is more variable for the Sunmaid brand, it would appear that some of the Sunmaid raisins are large while others are small. The individual sizes of the generic raisins are not as variable.

2.57 **a** The largest observation found in the data from Exercise 1.26 is 32.3, while the smallest is 0.2. Therefore the range is $R = 32.3 - 0.2 = 32.1$.

b Using the range, the approximate value for *s* is: $s \approx R/4 = 32.1/4 = 8.025$.

c Calculate $n = 50$, $\sum x_i = 418.4$ and $\sum x_i^2 = 6384.34$. Then

$$s = \sqrt{\frac{\sum x_i^2 - \frac{\left(\sum x_i\right)^2}{n}}{n-1}} = \sqrt{\frac{6384.34 - \frac{(418.4)^2}{50}}{49}} = 7.671$$

2.59 The ordered data are shown below.

0.2	2.0	4.3	8.2	14.7
0.2	2.1	4.4	8.3	16.7
0.3	2.4	5.6	8.7	18.0
0.4	2.4	5.8	9.0	18.0
1.0	2.7	6.1	9.6	18.4
1.2	3.3	6.6	9.9	19.2
1.3	3.5	6.9	11.4	23.1
1.4	3.7	7.4	12.6	24.0
1.6	3.9	7.4	13.5	26.7
1.6	4.1	8.2	14.1	32.3

Since $n = 50$, the position of the median is $0.5(n+1) = 25.5$ and the positions of the lower and upper quartiles are $0.25(n+1) = 12.75$ and $0.75(n+1) = 38.25$.

Then $m = (6.1 + 6.6)/2 = 6.35$, $Q_1 = 2.1 + 0.75(2.4 - 2.1) = 2.325$ and
$Q_3 = 12.6 + 0.25(13.5 - 12.6) = 12.825$. Then $IQR = 12.825 - 2.325 = 10.5$.

The *lower and upper fences* are:
$$Q_1 - 1.5 IQR = 2.325 - 15.75 = -13.425$$
$$Q_3 + 1.5 IQR = 12.825 + 15.75 = 28.575$$

and the box plot is shown below. There is one outlier, $x = 32.3$. The distribution is skewed to the right.

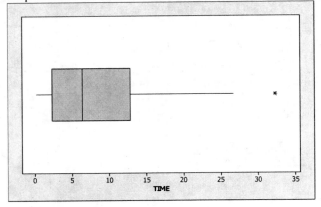

2.61   First calculate the intervals:
$$\overline{x} \pm s = 0.17 \pm 0.01 \qquad \text{or } 0.16 \text{ to } 0.18$$
$$\overline{x} \pm 2s = 0.17 \pm 0.02 \qquad \text{or } 0.15 \text{ to } 0.19$$
$$\overline{x} \pm 3s = 0.17 \pm 0.03 \qquad \text{or } 0.14 \text{ to } 0.20$$

a   If no prior information as to the shape of the distribution is available, we use Tchebysheff's Theorem. We would expect at least $(1 - 1/1^2) = 0$ of the measurements to fall in the interval 0.16 to 0.18; at least $(1 - 1/2^2) = 3/4$ of the measurements to fall in the interval 0.15 to 0.19; at least $(1 - 1/3^2) = 8/9$ of the measurements to fall in the interval 0.14 to 0.20.

b   According to the Empirical Rule, approximately 68% of the measurements will fall in the interval 0.16 to 0.18; approximately 95% of the measurements will fall between 0.15 to 0.19; approximately 99.7% of the measurements will fall between 0.14 and 0.20. Since mound-shaped distributions are so frequent, if we do have a sample size of 30 or greater, we expect the sample distribution to be mound-shaped. Therefore, in this exercise, we would expect the Empirical Rule to be suitable for describing the set of data.

c   If the chemist had used a sample size of four for this experiment, the distribution would not be mound-shaped. Any possible histogram we could construct would be non-mound-shaped. We can use at most 4 classes, each with frequency 1, and we will not obtain a histogram that is even close to mound-shaped. Therefore, the Empirical Rule would not be suitable for describing $n = 4$ measurements.

2.63   The following information is available:
$$n = 400, \ \overline{x} = 600, \ s^2 = 4900$$

The standard deviation of these scores is then 70, and the results of Tchebysheff's Theorem follow:

k	$\overline{x} \pm ks$	Interval	Tchebysheff
1	600 ± 70	530 to 670	at least 0
2	600 ± 140	460 to 740	at least 0.75
3	600 ± 210	390 to 810	at least 0.89

If the distribution of scores is mound-shaped, we use the Empirical Rule, and conclude that approximately 68% of the scores would lie in the interval 530 to 670 (which is $\bar{x} \pm s$). Approximately 95% of the scores would lie in the interval 460 to 740.

2.65 **a** Max = 27, Min = 20.2 and the range is $R = 27 - 20.2 = 6.8$.

b Answers will vary. A typical histogram is shown below. The distribution is slightly skewed to the left.

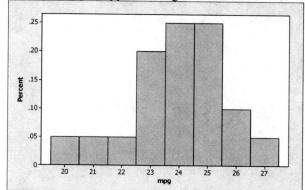

c Calculate $n = 20$, $\sum x_i = 479.2$, $\sum x_i^2 = 11532.82$. Then

$$\bar{x} = \frac{\sum x_i}{n} = \frac{479.2}{20} = 23.96$$

$$s = \sqrt{\frac{\sum x_i^2 - \frac{(\sum x_i)^2}{n}}{n-1}} = \sqrt{\frac{11532.82 - \frac{(479.2)^2}{20}}{19}} = \sqrt{2.694} = 1.641$$

d The sorted data is shown below:

```
20.2    21.3    22.2    22.7    22.9
23.1    23.2    23.6    23.7    24.2
24.4    24.4    24.6    24.7    24.7
24.9    25.3    25.9    26.2    27.0
```

The z-scores for $x = 20.2$ and $x = 27$ are

$$z = \frac{x - \bar{x}}{s} = \frac{20.2 - 23.96}{1.641} = -2.29 \text{ and } z = \frac{x - \bar{x}}{s} = \frac{27 - 23.96}{1.641} = 1.85$$

Since neither of the z-scores are greater than 3 in absolute value, the measurements are not judged to be outliers.

e The position of the median is $0.5(n+1) = 10.5$ and the median is $m = (24.2 + 24.4)/2 = 24.3$.

f The positions of the quartiles are $0.25(n+1) = 5.25$ and $0.75(n+1) = 15.75$. Then $Q_1 = 22.9 + 0.25(23.1 - 22.9) = 22.95$ and $Q_3 = 24.7 + 0.75(24.9 - 24.7) = 24.85$.

2.67 **a** The range is $R = 71 - 40 = 31$ and the range approximation is
$$s \approx R/4 = 31/4 = 7.75$$

b Calculate $n = 10$, $\sum x_i = 592$, $\sum x_i^2 = 36014$. Then

$$\bar{x} = \frac{\sum x_i}{n} = \frac{592}{10} = 59.2$$

$$s = \sqrt{\frac{\sum x_i^2 - \frac{(\sum x_i)^2}{n}}{n-1}} = \sqrt{\frac{36014 - \frac{(592)^2}{10}}{9}} = \sqrt{107.5111} = 10.369$$

27

The sample standard deviation calculated above is of the same order as the approximated value found in part **a**.

c The ordered set is:

$$40, 49, 52, 54, 59, 61, 67, 69, 70, 71$$

Since $n = 10$, the positions of m, Q_1, and Q_3 are 5.5, 2.75 and 8.25 respectively, and $m = (59 + 61)/2 = 60$, $Q_1 = 49 + 0.75(52 - 49) = 51.25$, $Q_3 = 69.75$ and $IQR = 69.75 - 51.25 = 18.5$.

The *lower and upper fences* are:

$$Q_1 - 1.5IQR = 51.25 - 27.75 = 23.5$$
$$Q_3 + 1.5IQR = 69.75 + 27.75 = 97.50$$

and the box plot is shown below. There are no outliers and the data set is slightly skewed left.

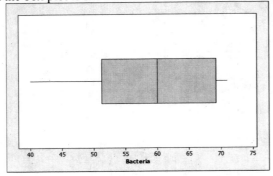

2.69 If the distribution is mound-shaped, then almost all of the measurements will fall in the interval $\mu \pm 3\sigma$, which is an interval 6σ in length. That is, the range of the measurements should be approximately 6σ. In this case, the range is $800 - 200 = 600$, so that $\sigma \approx 600/6 = 100$.

2.71 The stem lengths are approximately normal with mean 15 and standard deviation 2.5.
a In order to determine the percentage of roses with length less than 12.5, we must determine the proportion of the curve which lies within the shaded area in the figure below. Using the Empirical Rule, the proportion of the area between 12.5 and 15 is half of 0.68 or 0.34. Hence, the fraction below 12.5 would be $0.5 - 0.34 = 0.16$ or 16%.

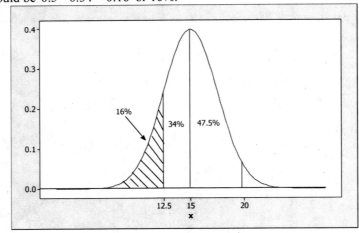

b Refer to the figure shown above. Again we use the Empirical Rule. The proportion of the area between 12.5 and 15 is half of 0.68 or 0.34, while the proportion of the area between 15 and 20 is half of 0.95 or 0.475. The total area between 12.5 and 20 is then $0.34 + 0.475 = .815$ or 81.5%.

28

2.73   The diameters of the trees are approximately mound-shaped with mean 14 and standard deviation 2.8.

   a   The value $x = 8.4$ lies two standard deviations below the mean, while the value $x = 22.4$ is three standard deviations above the mean. Use the Empirical Rule. The fraction of trees with diameters between 8.4 and 14 is half of 0.95 or 0.475, while the fraction of trees with diameters between 14 and 22.4 is half of 0.997 or 0.4985. The total fraction of trees with diameters between 8.4 and 22.4 is

$$0.475 + 0.4985 = .9735$$

   b   The value $x = 16.8$ lies one standard deviation above the mean. Using the Empirical Rule, the fraction of trees with diameters between 14 and 16.8 is half of 0.68 or 0.34, and the fraction of trees with diameters greater than 16.8 is

$$0.5 - 0.34 = .16$$

2.75   **a**   It is known that duration times are approximately normal, with mean 75 and standard deviation 20. In order to determine the probability that a commercial lasts less than 35 seconds, we must determine the fraction of the curve which lies within the shaded area in the figure below. Using the Empirical Rule, the fraction of the area between 35 and 75 is half of 0.95 or 0.475. Hence, the fraction below 35 would be $0.5 - 0.475 = 0.025$.

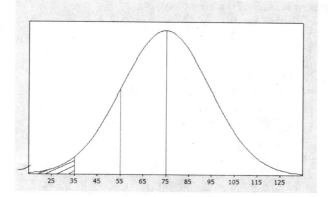

   b   The fraction of the curve area that lies above the 55 second mark may again be determined by using the Empirical Rule. Refer to the figure in part **a**. The fraction between 55 and 75 is 0.34 and the fraction above 75 is 0.5. Hence, the probability that a commercial lasts longer than 55 seconds is $0.5 + 0.34 = 0.84$.

2.77   **a**   The percentage of colleges that have between 145 and 205 teachers corresponds to the fraction of measurements expected to lie within two standard deviations of the mean. Tchebysheff's Theorem states that this fraction will be at least ¾ or 75%.
   b   If the population is normally distributed, the Empirical Rule is appropriate and the desired fraction is calculated. Referring to the normal distribution shown below, the fraction of area lying between 175 and 190 is 0.34, so that the fraction of colleges having more than 190 teachers is $0.5 - 0.34 = 0.16$.

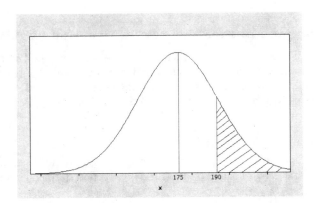

2.79 Notice that two (Sosa and McGuire) of the four players have relatively symmetric distributions. The whiskers are the same length and the median line is close to the middle of the box. The variability of the distributions is similar for all four players, but Barry Bonds has a distribution with a long right whisker, meaning that there may be an unusually large number of homers during one of his seasons. The distribution for Babe Ruth is slightly different from the others. The median line to the right of middle indicates a distribution skewed to the left; that there were a few seasons in which his homerun total was unusually low. In fact, the median number of homeruns for the other three players are all about 34-35, while Babe Ruth's median number of homeruns is closer to 40.

2.81 **a** Calculate $n = 50$, $\sum x_i = 418$, so that $\bar{x} = \dfrac{\sum x_i}{n} = \dfrac{418}{50} = 8.36$.

b The position of the median is $.5(n+1) = 25.5$ and $m = (4 + 4)/2 = 4$.

c Since the mean is larger than the median, the distribution is skewed to the right.

d Since $n = 50$, the positions of Q_1 and Q_3 are $.25(51) = 12.75$ and $.75(51) = 38.25$, respectively Then $Q_1 = 0 + 0.75(1-0) = 12.75$, $Q_3 = 17 + .25(19 - 17) = 17.5$ and $IQR = 17.5 - .75 = 16.75$.
The *lower and upper fences* are:
$$Q_1 - 1.5IQR = .75 - 25.125 = -24.375$$
$$Q_3 + 1.5IQR = 17.5 + 25.125 = 42.625$$

and the box plot is shown below. There are no outliers and the data is skewed to the right.

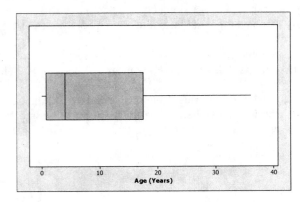

2.83 Answers will vary. Students should notice that the distribution of baseline measurements is relatively mound-shaped. Therefore, the Empirical Rule will provide a very good description of the data. A measurement which is further than two or three standard deviations from the mean would be considered unusual.

30

2.85 **a** For $n = 25$, the position of the median is $0.5(n+1) = 13$ and the positions of the quartiles are $0.25(n+1) = 6.5$ and $0.75(n+1) = 19.5$. Then $m = 4.2$, $Q_1 = (3.7 + 3.8)/2 = 3.75$ and $Q_3 = (4.7 + 4.8)/2 = 4.75$.

Then the five-number summary is

Min	Q_1	Median	Q_3	Max
2.5	3.75	4.2	4.75	5.7

b-c Calculate $IQR = Q_3 - Q_1 = 4.75 - 3.75 = 1$. Then the *lower and upper fences* are:

$$Q_1 - 1.5 IQR = 3.75 - 1.5 = 2.25$$
$$Q_3 + 1.5 IQR = 4.75 + 1.5 = 6.25$$

There are no unusual measurements, and the box plot is shown on the next page.

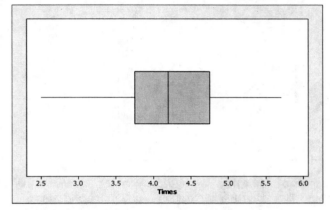

d Answers will vary. A stem and leaf plot, generated by *Minitab,* is shown below. The data is roughly mound-shaped.

Stem-and-Leaf Display: Times

```
Stem-and-leaf of Times   N  = 25
Leaf Unit = 0.10

   1    2  5
   4    3  013
  10    3  678899
  (7)   4  1222334
   8    4  7788
   4    5  234
   1    5  7
```

3: Describing Bivariate Data

3.1   **a**   The side-by-side pie charts are constructed as in Chapter 1 for each of the two groups (men and women) and are displayed below using the percentages shown in the table below.

	Group 1	Group 2	Group 3	Total
Men	23%	31%	46%	100%
Women	8%	57%	35%	100%

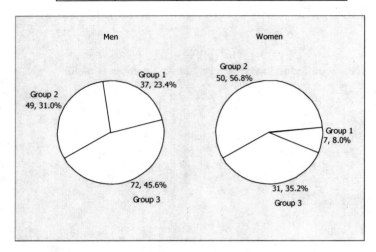

    b-c   The side-by-side and stacked bar charts in the next two figures measure the frequency of occurrence for each of the three groups. A separate bar (or portion of a bar) is used for men and women.

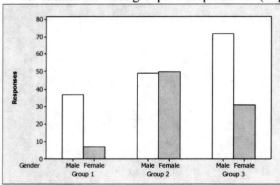 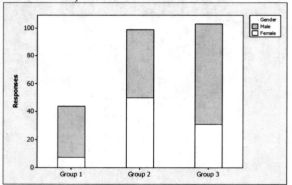

    d   The differences in the proportions of men and women in the three groups is most graphically portrayed by the pie charts, since the unequal number of men and women tend to confuse the interpretation of the bar charts. However, the bar charts are useful in retaining the actual frequencies of occurrence in each group, which is lost in the pie chart.

3.3   **a**   Similar to Exercise 3.1. Any of the comparative charts (side-by-side pie charts, stacked or side-by-side bar charts) can be used.
    b-c   The two types of comparative bar charts are shown on the next page. The amounts spent in each of the four categories seem to be quite different for men and women, except in category C. In category C which involves the largest dollar amount of purchase, there is little difference between the genders.

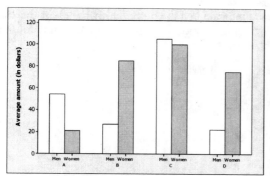
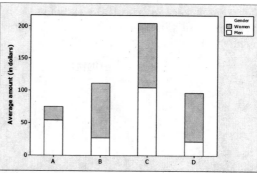

d   Although it is really a matter of preference, the only advantage to the stacked chart is that the reader can easily see the total dollar amount for each category. For comparison purposes, the side-by-side chart may be better.

3.5   **a**   The population of interest is the population of responses to the question about free time for all parents and children in the United States. The sample is the set of responses generated for the 198 parents and 200 children in the survey.

b   The data can be considered bivariate if, for each person interviewed, we record the person's relationship (Parent or Child) and their response to the question (just the right amount, not enough, too much, don't know). Since the measurements are not numerical in nature, the variables are qualitative.

c   The entry in a cell represents the number of people who fell into that relationship-opinion category.

d   A pie chart is created for both the "parent" and the "children" categories. The size of each sector angle is proportional to the fraction of measurements falling into that category.

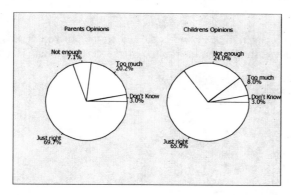

e   Either stacked or comparative bar charts could be used, but since the height of the bar represents the frequency of occurrence (and hence is tied to the sample size), this type of chart would be misleading. The comparative pie charts are the best choice.

3.7   **a**   The side-by-side bar chart is shown below.

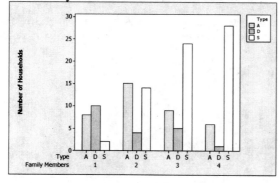

33

b-c The stacked bar chart is shown below. Both charts indicate that the more family members there are, the more likely it is that the family lives in a duplex or a single residence. The fewer the number of family members, the more likely it is that the family lives in an apartment.

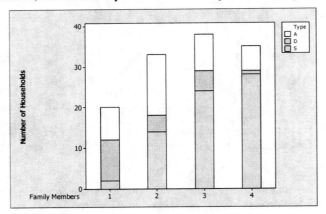

3.9 **a** The change in *y* for a one-unit change in *x* is defined as the *slope* of the line—the coefficient of the *x* variable. For this line, the slope is $b = 0.5$.

b Since the slope of the line is positive, the values of *y* will *increase* as the values of *x* increase.

c The point at which the line crosses the *y*-axis is called the *y*-intercept, and is the constant term in the equation of the line. For this line, the *y*-intercept is $a = 2.0$.

d When $x = 2.5$, the predicted value of *y* is $y = 2.0 + 0.5(2.5) = 3.25$. When $x = 4.0$, the predicted value of *y* is $y = 2.0 + 0.5(4.0) = 4.0$.

3.11 **a** The first variable (*x*) is the first number in the pair and is plotted on the horizontal axis, while the second variable (*y*) is the second number in the pair and is plotted on the vertical axis. The scatterplot is shown in the figure below.

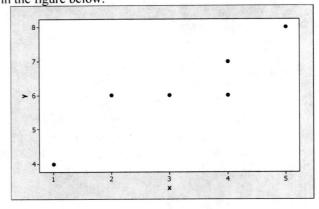

b There appears to be a positive relationship between *x* and *y*; that is, as *x* increases, so does *y*.

c Use your scientific calculator to calculate the sums, sums of square sand sum of cross products for the pairs (x_i, y_i).

$$\sum x_i = 19; \ \sum y_i = 37; \ \sum x_i^2 = 71; \ \sum y_i^2 = 237; \ \sum x_i y_i = 126$$

Then the covariance is

$$s_{xy} = \frac{\sum x_i y_i - \frac{(\sum x_i)(\sum y_i)}{n}}{n-1} = \frac{126 - \frac{(19)37}{6}}{5} = 1.76667$$

and the sample standard deviations are

34

$$s_x = \sqrt{\frac{\sum x_i - \frac{\left(\sum x_i\right)^2}{n}}{n-1}} = \sqrt{\frac{71 - \frac{(19)^2}{6}}{5}} = 1.472 \text{ and } s_y = \sqrt{\frac{\sum y_i - \frac{\left(\sum y_i\right)^2}{n}}{n-1}} = \sqrt{\frac{237 - \frac{(37)^2}{6}}{5}} = 1.329$$

The correlation coefficient is

$$r = \frac{s_{xy}}{s_x s_y} = \frac{1.76667}{(1.472)(1.329)} = 0.902986 \approx 0.903$$

d   The slope and y-intercept of the regression line are

$$b = r\frac{s_y}{s_x} = 0.902986\left(\frac{1.329}{1.472}\right) = 0.81526 \text{ and } a = \bar{y} - b\bar{x} = \frac{37}{6} - 0.81526\left(\frac{19}{6}\right) = 3.58$$

and the equation of the regression line is $y = 3.58 + 0.815x$.

The graph of the data points and the best fitting line is shown below. The line fits through the data points.

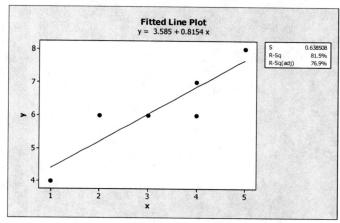

3.13   **a**   Similar to Exercise 3.11. The scatterplot is shown below.

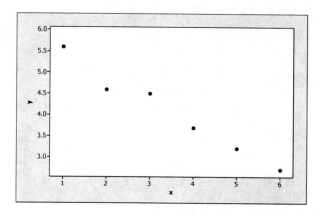

b   There appears to be a negative relationship between x and y; that is, as x increase, y decreases.

c   Use your scientific calculator to calculate the sums, sums of squares and sum of cross products for the pairs (x_i, y_i).

$$\sum x_i = 21; \ \sum y_i = 24.3; \ \sum x_i^2 = 91; \ \sum y_i^2 = 103.99; \ \sum x_i y_i = 75.3$$

Then the covariance is

$$s_{xy} = \frac{\sum x_i y_i - \dfrac{(\sum x_i)(\sum y_i)}{n}}{n-1} = \frac{75.3 - \dfrac{(21)24.3}{6}}{5} = -1.95$$

and the sample standard deviations are

$$s_x = \sqrt{\frac{\sum x_i - \dfrac{(\sum x_i)^2}{n}}{n-1}} = \sqrt{\frac{91 - \dfrac{(21)^2}{6}}{5}} = 1.8708 \text{ and}$$

$$s_y = \sqrt{\frac{\sum y_i - \dfrac{(\sum y_i)^2}{n}}{n-1}} = \sqrt{\frac{103.99 - \dfrac{(24.3)^2}{6}}{5}} = 1.0559$$

The correlation coefficient is

$$r = \frac{s_{xy}}{s_x s_y} = \frac{-1.95}{(1.8708)(1.0559)} = -0.987$$

This value of r indicates a strong negative relationship between x and y.

3.15   **a**   Use your scientific calculator. You can verify that $\sum x_i = 17$; $\sum y_i = 741.67$; $\sum x_i^2 = 59$; $\sum y_i^2 = 98084.2579$; $\sum x_i y_i = 2359.51$. Then the covariance is

$$s_{xy} = \frac{\sum x_i y_i - \dfrac{(\sum x_i)(\sum y_i)}{n}}{n-1} = \frac{2359.51 - \dfrac{17(741.67)}{6}}{5} = 51.62233$$

The sample standard deviations are $s_x = 1.47196$ and $s_y = 35.79160$ so that $r = 0.9799$. Then

$$b = r\frac{s_y}{s_x} = 23.8257 \text{ and } a = \bar{y} - b\bar{x} = 123.611667 - 23.8257(2.8333) = 56.106$$

and the equation of the regression line is $y = 56.106 + 23.826x$.

  b   The graph of the data points and the best fitting line is shown below.

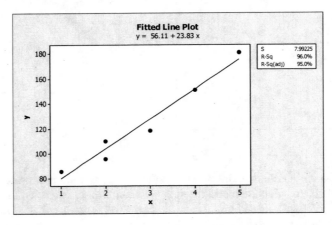

  c   When $x = 6$, the estimated value of y is $y = 56.106 + 23.826(6) = 199.06$. However, it is risky to try to estimate the value of y for a value of x outside of the experimental region – that is, the range of x values for which you have collected data.

3.17   **a-b** The scatterplot is shown on the next page. There is a slight positive trend between pre- and post-test scores, but the trend is not too pronounced.

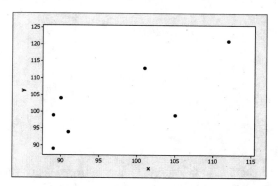

c   Calculate $n = 7$; $\sum x_i = 677$; $\sum y_i = 719$; $\sum x_i^2 = 65{,}993$; $\sum y_i^2 = 74{,}585$; $\sum x_i y_i = 70{,}006$. Then the covariance is

$$s_{xy} = \frac{\sum x_i y_i - \dfrac{(\sum x_i)(\sum y_i)}{n}}{n-1} = 78.071429$$

The sample standard deviations are $s_x = 9.286447$ and $s_y = 11.056134$ so that $r = 0.760$. This is a relatively strong positive correlation, confirming the interpretation of the scatterplot.

3.19   **a**   Since we would be interested in predicting the price of a LCD TV based on its size, the price is the dependent variable (y) and size is the independent variable (x).
b   The scatterplot is shown below. The relationship is does not appear to be linear; in fact, it's hard to see much of a relationship at all.

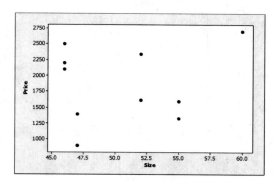

3.21   **a-b** The data is graphed as a scatterplot in the figure below, with the time in months plotted on the horizontal axis and the number of books on the vertical axis. The data points are then connected to form a line graph. There is a very distinct pattern in this data, with the number of books increasing with time, a response which might be modeled by a quadratic equation. The professor's productivity appears to increase, with less time required to write later books.

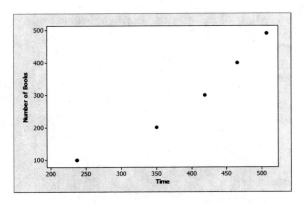

3.23   **a**   The following variables have been considered in this survey: the number of users (quantitative), the year (2009 or 2010, qualitative), the education level (qualitative) and the city (qualitative).
b   The population of interest is the population of responses for all *Facebook* users in 2009 and 2010. This is a fixed population at whatever moment in time it happened to be measured.
c   Comparative (side-by-side) bar charts have been used. An alternative presentation can be obtained by using comparative pie charts or stacked bar charts.
d   Answers will vary. It seems that all of the cities experienced an increase in users from 2009 to 2010, with the largest   increases in older users (alumni or unknown).

3.25   The scatterplot for these two quantitative variables is shown below. Notice the almost perfect positive correlation. The correlation coefficient and best fitting line can be calculated for descriptive purposes.

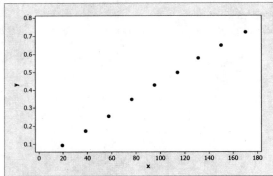

Calculate $n = 9$; $\sum x_i = 850.8$; $\sum y_i = 3.755$; $\sum x_i^2 = 101,495.78$; $\sum y_i^2 = 1.941467$; $\sum x_i y_i = 443.7727$.
Then the covariance is

$$s_{xy} = \frac{\sum x_i y_i - \dfrac{(\sum x_i)(\sum y_i)}{n}}{n-1} = 11.10000417$$

The sample standard deviations are $s_x = 51.31620$ and $s_y = 0.216448$ so that $r = 0.999$. Then

$$b = r\frac{s_y}{s_x} = 0.004215 \text{ and } a = \bar{y} - b\bar{x} = 0.417222 - 0.004215(94.5333) = 0.0187$$

and the equation of the regression line is $y = 0.0187 + 0.0042x$.

3.27   **a**   Using the *Minitab* output, calculate

$$r = \frac{s_{xy}}{s_x s_y} = \frac{278.0}{\sqrt{(6.8)(12839.1)}} = 0.941$$

(Using the original data, the student will obtain $r = .944$.)

38

b Since the number of weeks in release would tend to explain the gross to date, we could consider $x =$ number of weeks in release to be the independent variable, and $y =$ gross to date to be the dependent variable.

c Using the *Minitab* output, calculate

$$r_{GT} = \frac{s_{xy}}{s_x s_y} = \frac{2550.7}{\sqrt{(19.6)(550521.6)}} = 0.777 \quad \text{and} \quad r_{T/PT} = \frac{s_{xy}}{s_x s_y} = \frac{421558.5}{\sqrt{(550521.6)(1700161.7)}} = 0.436$$

d Answers will vary from student to student.

3.29 **a-c** No. There seems to be a large cluster of points in the lower left hand corner showing no apparent relationship between the variables, while 7-10 data points from top left to bottom right show a negative linear relationship.

b The pattern described in parts **a** and **c** would indicate a weak correlation:

$$r = \frac{s_{xy}}{s_x s_y} = \frac{-72.176}{\sqrt{(9346.603)(702.776)}} = -0.028$$

d Number of waste sites is only slightly affected by the size of the state. Some other possible explanatory variables might be local environmental regulations, population per square mile, or geographical region in the United States.

3.31 **a** The variables measured are amount of aluminum oxide (quantitative), and archeological site (qualitative).

b Looking at the box plots, you can see that there are higher levels of aluminum oxide at the Ashley Rails and Island Thorns sites. The variability is about the same at the three sites.

3.33 **a-b** The scatterplot is shown below. The relationship appears to be positive linear, but fairly weak.

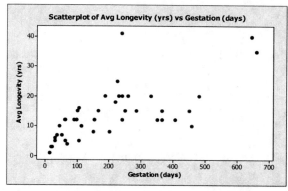

c There is one outlier to the linear trend [hippo-(238, 41)] and two data points that are unusually large [African elephant-(660, 35) and Asian elephant-(645, 40)].

d When the three data points from part c are removed, the scatterplot is as shown below. It appears that the relationship may be more curvilinear than linear.

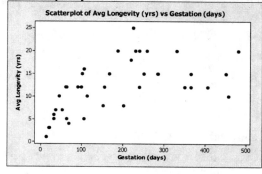

3.35 **a** The scatterplot is shown below. There is a positive linear relationship between armspan and height.

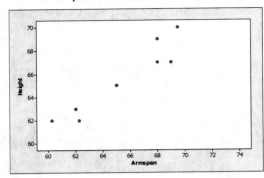

b Calculate $n = 8$; $\sum x_i = 524$; $\sum y_i = 525$; $\sum x_i^2 = 34413.375$; $\sum y_i^2 = 34521$; $\sum x_i y_i = 34462$.
Then the covariance is

$$s_{xy} = \frac{\sum x_i y_i - \dfrac{(\sum x_i)(\sum y_i)}{n}}{n-1} = 10.64286$$

The sample standard deviations are $s_x = 3.61297$ and $s_y = 3.1139089$ so that $r = 0.946$.

c Since DaVinci indicated that a person's armspan is roughly equal to his height, the slope of the line should be approximately equal to 1.

d Calculate (using full accuracy)

$$b = r\frac{s_y}{s_x} = .815 \text{ and } a = \bar{y} - b\bar{x} = 65.625 - .815(65.5) = 12.22$$

and the equation of the regression line is $y = 12.22 + .815x$.

3.37 **a** Calculate $n = 8$; $\sum x_i = 575$; $\sum y_i = 558$; $\sum x_i^2 = 43,163$; $\sum y_i^2 = 42,196$; $\sum x_i y_i = 41,664$. Then the covariance is

$$s_{xy} = \frac{\sum x_i y_i - \dfrac{(\sum x_i)(\sum y_i)}{n}}{n-1} = 222.535714$$

The sample standard deviations are $s_x = 16.190275$ and $s_y = 21.631657$ so that $r = 0.635$.

b Calculate (using full accuracy)

$$b = r\frac{s_y}{s_x} = .848968 \text{ and } a = \bar{y} - b\bar{x} = 69.75 - .848968(71.875) = 8.730$$

and the equation of the regression line is $y = 8.730 + .849x$.

c When $x = 85$, the predicted midterm 2 score is $y = 8.730 + .849(85) = 80.895$.

3.39 **a** Calculate $n = 8$; $\sum x_i = 634$; $\sum y_i = 386$; $\sum x_i^2 = 52270$; $\sum y_i^2 = 19876$; $\sum x_i y_i = 32136$.
Then the covariance is

$$s_{xy} = \frac{\sum x_i y_i - \dfrac{(\sum x_i)(\sum y_i)}{n}}{n-1} = 220.78571$$

The sample standard deviations are $s_x = 17.010501$ and $s_y = 13.3710775$ so that $r = 0.971$.

b Since the correlation coefficient is so close to 1, the strong correlation indicates that the second and quicker test could be used in place of the longer test-interview.

40

3.41 **a** The scatterplot is shown below. The relationship is positive linear.

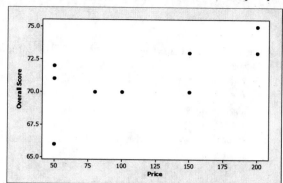

b Calculate $n = 9$; $\sum x_i = 1030$; $\sum y_i = 640$; $\sum x_i^2 = 148,900$; $\sum y_i^2 = 45,564$; $\sum x_i y_i = 74,100$. Then the covariance is

$$s_{xy} = \frac{\sum x_i y_i - \dfrac{(\sum x_i)(\sum y_i)}{n}}{n-1} = 106.9444444$$

The sample standard deviations are $s_x = 62.271806$ and $s_y = 2.5712081$ so that $r = .668$.

c Calculate (using full accuracy)

$$b = r\frac{s_y}{s_x} = .027579 \text{ and } a = \overline{y} - b\overline{x} = 71.11111 - .027579(114.44444) = 67.9548$$

and the equation of the regression line is $y = 67.955 + .028x$.

4: Probability and Probability Distributions

4.1　**a**　This experiment involves tossing a single die and observing the outcome. The sample space for this experiment consists of the following simple events:

E_1: Observe a 1　　　　　E_4: Observe a 4
E_2: Observe a 2　　　　　E_5: Observe a 5
E_3: Observe a 3　　　　　E_6: Observe a 6

b　Events A through F are compound events and are composed in the following manner:

A: (E_2)　　　　　　　　D: (E_2)
B: (E_2, E_4, E_6)　　　　E: (E_2, E_4, E_6)
C: (E_3, E_4, E_5, E_6)　　F: contains no simple events

c　Since the simple events E_i, $i = 1, 2, 3, ..., 6$ are equally likely, $P(E_i) = 1/6$.

d　To find the probability of an event, we sum the probabilities assigned to the simple events in that event. For example,

$$P(A) = P(E_2) = \frac{1}{6}$$

Similarly, $P(D) = 1/6; P(B) = P(E) = P(E_2) + P(E_4) + P(E_6) = \frac{3}{6} = \frac{1}{2}$; and $P(C) = \frac{4}{6} = \frac{2}{3}$. Since event F contains no simple events, $P(F) = 0$.

4.3　It is given that $P(E_1) = .45$ and that $3P(E_2) = .45$, so that $P(E_2) = .15$. Since $\sum_S P(E_i) = 1$, the remaining 8 simple events must have probabilities whose sum is $P(E_3) + P(E_4) + ... + P(E_{10}) = 1 - .45 - .15 = .4$. Since it is given that they are equiprobable,

$$P(E_i) = \frac{.4}{8} = .05 \text{ for } i = 3, 4, ..., 10$$

4.5　**a**　The experiment consists of choosing three coins at random from four. The order in which the coins are drawn is unimportant. Hence, each simple event consists of a triplet, indicating the three coins drawn. Using the letters N, D, Q, and H to represent the nickel, dime, quarter, and half-dollar, respectively, the four possible simple events are listed below.

E_1: (NDQ)　　　E_2: (NDH)　　　E_3: (NQH)　　　E_4: (DQH)

b　The event that a half-dollar is chosen is associated with the simple events E_2, E_3, and E_4. Hence,

$$P[\text{choose a half-dollar}] = P(E_2) + P(E_3) + P(E_4) = \frac{1}{4} + \frac{1}{4} + \frac{1}{4} = \frac{3}{4}$$

since each simple event is equally likely.

c　The simple events along with their monetary values follow:

E_1　NDQ　$0.40
E_2　NDH　0.65
E_3　NQH　0.80
E_4　DQH　0.85

Hence, $P[\text{total amount is \$0.60 or more}] = P(E_2) + P(E_3) + P(E_4) = 3/4$.

4.7　Label the five balls as R_1, R_2, R_3, Y_1 and Y_2. The selection of two balls is accomplished in two stages to produce the simple events in the tree diagram on the next page.

First Ball	Second Ball	Simple Events		First Ball	Second Ball	Simple Events
	R_2	R_1R_2			R_1	Y_1R_1
	R_3	R_1R_3			R_2	Y_1R_2
R_1	Y_1	R_1Y_1		Y_1	R_3	Y_1R_3
	Y_2	R_1Y_2			Y_2	Y_1Y_2
	R_1	R_2R_1			R_1	Y_2R_1
R_2	R_3	R_2R_3		Y_2	R_2	Y_2R_2
	Y_1	R_2Y_1			R_3	Y_2R_3
	Y_2	R_2Y_2			Y_1	Y_2Y_1
	R_1	R_3R_1				
R_3	R_2	R_3R_2				
	Y_1	R_3Y_1				
	Y_2	R_3Y_2				

4.9 The four possible outcomes of the experiment, or simple events, are represented as the cells of a 2×2 table, and have probabilities as given in the table.

    a P[adult judged to need glasses] $= .44 + .14 = .58$

    b P[adult needs glasses but does not use them] $= .14$

    c P[adult uses glasses] $= .44 + .02 = .46$

4.11   **a**   *Experiment*: Select three people and record their gender (M or F).

    b   Extend the tree diagram in Figure 4.3 of the text to include one more coin toss (a total of $n = 3$). Then replace the H and T by M and F to obtain the 8 possible simple events shown below:

            FFF   FMM   MFM   MMF
            MFF   FMF   FFM   MMM

    c   Since there are $N = 8$ equally likely simple events, each is assigned probability, $P(E_i) = 1/N = 1/8$.

    d-e   Sum the probabilities of the appropriate simple events:

$$P(\text{only one man}) = P(MFF) + P(FMF) + P(FFM) = 3\left(\frac{1}{8}\right) = \frac{3}{8}$$

$$P(\text{all three are women}) = P(FFF) = \frac{1}{8}$$

4.13   **a**   *Experiment*: A taster tastes and ranks three varieties of tea A, B, and C, according to preference.

    b   Simple events in S are in triplet form.

        $E_1 : (A, B, C)$ $E_4 : (B, C, A)$

        $E_2 : (A, C, B)$ $E_5 : (C, B, A)$

        $E_3 : (B, A, C)$ $E_6 : (C, A, B)$

Here the most desirable is in the first position, the next most desirable is in the second position, and the least desirable is in third position.

    c   Define the events   D: variety A is ranked first
                                          F: variety A is ranked third

Then

$$P(D) = P(E_1) + P(E_2) = 1/6 + 1/6 = 1/3$$

The probability that A is least desirable is

$$P(F) = P(E_5) + P(E_6) = 1/6 + 1/6 = 1/3$$

4.15 Similar to Exercise 4.9. The four possible outcomes of the experiment, or simple events, are represented as the cells of a 2×2 table, and have probabilities (when divided by 300) as given in the table.

    a $P(\text{normal eyes and normal wing size}) = 140/300 = .467$

    b $P(\text{vermillion eyes}) = (3 + 151)/300 = 154/300 = .513$

    c $P(\text{either vermillion eyes or miniature wings or both}) = (3 + 151 + 6)/300 = 160/300 = .533$

43

4.17   Use the *mn* Rule. There are $10(8) = 80$ possible pairs.

4.19   **a** $P_3^5 = \dfrac{5!}{2!} = 5(4)(3) = 60$ **b** $P_9^{10} = \dfrac{10!}{1!} = 3,628,800$

c $P_6^6 = \dfrac{6!}{0!} = 6! = 720$ **d** $P_1^{20} = \dfrac{20!}{19!} = 20$

4.21   Since order is important, you use *permutations* and $P_5^8 = \dfrac{8!}{3!} = 8(7)(6)(5)(4) = 6720$.

4.23   Use the extended *mn* Rule. The first die can fall in one of 6 ways, *and* the second and third die can each fall in one of 6 ways. The total number of simple events is $6(6)(6) = 216$.

4.25   Since order is unimportant, you use *combinations* and $C_3^{10} = \dfrac{10!}{3!7!} = \dfrac{10(9)(8)}{3(2)(1)} = 120$.

4.27   This exercise involves the arrangement of 6 different cities in all possible orders. Each city will be visited once and only once. Hence, order is important and elements are being chosen from a single set. Permutations are used and the number of arrangements is

$$P_6^6 = \dfrac{6!}{0!} = 6! = 6(5)(4)(3)(2)(1) = 720$$

4.29   **a**   Each student has a choice of 52 cards, since the cards are replaced between selections. The *mn* Rule allows you to find the total number of configurations for three students as $52(52)(52) = 140,608$.

b   Now each student must pick a different card. That is, the first student has 52 choices, but the second and third students have only 51 and 50 choices, respectively. The total number of configurations is found using the *mn* Rule or the rule for permutations:

$$mnt = 52(51)(50) = 132,600 \quad \text{or} \quad P_3^{52} = \dfrac{52!}{49!} = 132,600 .$$

c   Let A be the event of interest. Since there are 52 different cards in the deck, there are 52 configurations in which all three students pick the same card (one for each card). That is, there are $n_A = 52$ ways for the event A to occur, out of a total of $N = 140,608$ possible configurations from part **a**. The probability of interest is

$$P(A) = \dfrac{n_A}{N} = \dfrac{52}{140,608} = .00037$$

d   Again, let A be the event of interest. There are $n_A = 132,600$ ways (from part **b**) for the event A to occur, out of a total of $N = 140,608$ possible configurations from part **a**, and the probability of interest is

$$P(A) = \dfrac{n_A}{N} = \dfrac{132,600}{140,608} = .943$$

4.31   **a**   Since the order of selection for the five-card hand is unimportant, use *combinations* to find the number of possible hands as $N = C_5^{52} = \dfrac{52!}{5!47!} = \dfrac{52(51)(50)(49)(48)}{5(4)(3)(2)(1)} = 2,598,960$.

b   Since there are only four different suits, there are $n_A = 4$ ways to get a royal flush.

c   From parts **a** and **b**,

$$P(\text{royal flush}) = \dfrac{n_A}{N} = \dfrac{4}{2,598,960} = .000001539$$

4.33   Notice that a sample of 10 nurses will be the same no matter in which order they were selected. Hence, order is unimportant and combinations are used. The number of samples of 10 selected from a total of 90 is

$$C_{10}^{90} = \dfrac{90!}{10!80!} = \dfrac{2.0759076\left(10^{19}\right)}{3.6288\left(10^6\right)} = 5.720645\left(10^{12}\right)$$

4.35   **a**   Use the *mn* Rule. The Western conference team can be chosen in one of $m = 6$ ways, and there are 6 ways to choose the Eastern conference team, for a total of $N = mn = 6(6) = 36$ possible pairings.

b   You must choose Los Angeles from the first group and New York from the second group, so that $n_A = (1)(1) = 1$ and the probability is $n_A /N = 1/36$.

c   Since there is only one California team in the Western conference, there are five choices for the first team and six choices for the second team. Hence $n_A = 5(6) = 30$ and the probability is $n_A /N = 30/36 = 5/6$.

4.37   The situation presented here is analogous to drawing 5 items from a jar (the five members voting in favor of the plaintiff). If the jar contains 5 red and 3 white items (5 women and 3 men), what is the probability that all five items are red? That is, if there is no sex bias, five of the eight members are randomly chosen to be those voting for the plaintiff. What is the probability that all five are women? There are

$$N = C_5^8 = \frac{8!}{5!3!} = 56$$

simple events in the experiment, only one of which results in choosing 5 women. Hence,

$$P(\text{five women}) = \frac{1}{56}.$$

4.39   The monkey can place the twelve blocks in any order. Each arrangement will yield a simple event and hence the total number of simple events (arrangements) is $P_{12}^{12} = 12!$ It is necessary to determine the number of simple events in the event of interest (that he draws three of each kind, in order). First, he may draw the four different <u>types</u> of blocks in any order. Thus we need the number of ways of arranging these four items, which is $P_4^4 = 4!$ Once this order has been chosen, the three squares can be arranged in $P_3^3 = 3!$ ways, the three triangles can be arranged in $P_3^3 = 3!$ ways, and so on. Thus the total number of simple events in the event of interest is $P_4^4 \left(P_3^3 \right)^4$ and the associated probability is

$$\frac{P_4^4 \left(P_3^3 \right)^4}{P_{12}^{12}} = \frac{4!(3!)^4}{12!}$$

4.41   **a** $P\left(A^C \right) = 1 - P(A) = 1 - \dfrac{2}{5} = \dfrac{3}{5}$

b $P\left(A \cap B \right)^C = 1 - P(A \cup B) = 1 - \dfrac{1}{5} = \dfrac{4}{5}$

4.43   **a** $P(A \cup B) = P(A) + P(B) - P(A \cap B) = 2/5 + 4/5 - 1/5 = 5/5 = 1$

b $P(A \cap B) = P(A \mid B)P(B) = (1/4)(4/5) = 1/5$

c $P(B \cap C) = P(B \mid C)P(C) = (1/2)(2/5) = 1/5$

4.45   **a** $P(A \cap B) = P(B)P(A \mid B) = .5(.1) = .05$

b   Since $P(A) = P(A \mid B) = .1$, events A and B are independent.

c   Since $P(A \cap B) = 0$ and $P(A)P(B) = .1(.5) = .05$, $P(A \cap B) \neq P(A)P(B)$. Hence, events A and B are not independent.

d   Using the Addition Rule, we have
$$P(A \cup B) = P(A) + P(B) - P(A \cap B)$$
$$.65 = .6 - P(A \cap B) \text{ so that } P(A \cap B) = .05.$$

Since $P(A \cap B) \neq 0$, events A and B are not mutually exclusive.

4.47 **a** From Exercise 4.46, $P(A \cap B) = 1/3$, $P(A|B) = 1$, $P(A) = 1/2$, $P(A|B) \neq P(A)$, so that A and B are not independent. $P(A \cap B) \neq 0$, so that A and B are not mutually exclusive.

  b $P(A|C) = P(A \cap C)/P(C) = 0$, $P(A) = 1/2$, $P(A \cap C) = 0$. Since $P(A|C) \neq P(A)$, A and C are dependent. Since $P(A \cap C) = 0$, A and C are mutually exclusive.

4.49 **a** Since A and B are independent, $P(A \cap B) = P(A)P(B) = .4(.2) = .08$.

  b $P(A \cup B) = P(A) + P(B) - P(A \cap B) = .4 + .2 - (.4)(.2) = .52$

4.51 **a** Use the definition of conditional probability to find

$$P(B|A) = \frac{P(A \cap B)}{P(A)} = \frac{.12}{.4} = .3$$

  b Since $P(A \cap B) \neq 0$, A and B are not mutually exclusive.

  c If $P(B) = .3$, then $P(B) = P(B|A)$ which means that A and B are independent.

4.53 **a** From Exercise 4.52, since $P(A \cap B) = .34$, the two events are not mutually exclusive.

  b From Exercise 4.52, $P(A|B) = .425$ and $P(A) = .49$. The two events are not independent.

4.55 Define the following events:
  A: project is approved for funding
  D: project is disapproved for funding
For the first panel, $P(A_1) = .2$ and $P(D_1) = .8$. For the second panel, $P[\text{same decision as first panel}] = .7$ and $P[\text{reversal}] = .3$. That is,

$$P(A_2 | A_1) = P(D_2 | D_1) = .7 \text{ and } P(A_2 | D_1) = P(D_2 | A_1) = .3.$$

  a $P(A_1 \cap A_2) = P(A_1)P(A_2 | A_1) = .2(.7) = .14$

  b $P(D_1 \cap D_2) = P(D_1)P(D_2 | D_1) = .8(.7) = .56$

  c $P(D_1 \cap A_2) + P(A_1 \cap D_2) = P(D_1)P(A_2 | D_1) + P(A_1)P(D_2 | A_1) = .8(.3) + .2(.3) = .30$

4.57 Refer to Exercise 4.56.

  a From the table, $P(A \cap B) = .1$ while $P(A)P(B|A) = (.4)(.1/.4) = .1$

  b From the table, $P(A \cap B) = .1$ while $P(B)P(A|B) = (.37)(.1/.37) = .1$

  c From the table, $P(A \cup B) = .1 + .27 + .30 = .67$ while $P(A) + P(B) - P(A \cap B) = .4 + .37 - .10 = .67$.

4.59 Fix the birth date of the first person entering the room. Then define the following events:
  A_2: second person's birthday differs from the first
  A_3: third person's birthday differs from the first and second
  A_4: fourth person's birthday differs from all preceding
  $\vdots$
  A_n: n^{th} person's birthday differs from all preceding
Then

$$P(A) = P(A_2)P(A_3) \cdots P(A_n) = \left(\frac{364}{365}\right)\left(\frac{363}{365}\right) \cdots \left(\frac{365 - n + 1}{365}\right)$$

since at each step, one less birth date is available for selection. Since event B is the complement of event A,
$$P(B) = 1 - P(A)$$

  a For $n = 3$, $P(A) = \frac{(364)(363)}{(365)^2} = .9918$ and $P(B) = 1 - .9918 = .0082$.

b For $n = 4$, $P(A) = \dfrac{(364)(363)(362)}{(365)^3} = .9836$ and $P(B) = 1 - .9836 = .0164$

4.61 Let events A and B be defined as follows:
   A: article gets by the first inspector
   B: article gets by the second inspector
The event of interest is then the event $A \cap B$, that the article gets by both inspectors. It is given that $P(A) = .1$, and also that $P(B \mid A) = .5$. Applying the Multiplication Rule,
$$P(A \cap B) = P(A)P(B \mid A) = (.1)(.5) = .05$$

4.63 Define
   A: smoke is detected by device A
   B: smoke is detected by device B
If it is given that $P(A) = .95$, $P(B) = .98$, and $P(A \cap B) = .94$.

a $P(A \cup B) = P(A) + P(B) - P(A \cap B) = .95 + .98 - .94 = .99$

b $P(A^C \cap B^C) = 1 - P(A \cup B) = 1 - .99 = .01$

4.65 The two-way table in the text gives event counts for events A, B, C, P and G in the column and row marked "Totals". The interior of the table contains the six two-way intersection counts as shown below.

$A \cap P$	$A \cap G$
$B \cap P$	$B \cap G$
$C \cap P$	$C \cap G$

The necessary probabilities can be found using various rules of probability if not directly from the table.

a $P(A) = \dfrac{154}{256}$

b $P(G) = \dfrac{155}{256}$

c $P(A \cap G) = \dfrac{88}{256}$

d $P(G \mid A) = \dfrac{P(G \cap A)}{P(A)} = \dfrac{88/256}{154/256} = \dfrac{88}{154}$

e $P(G \mid B) = \dfrac{P(G \cap B)}{P(B)} = \dfrac{44/256}{67/256} = \dfrac{44}{67}$

f $P(G \mid C) = \dfrac{P(G \cap C)}{P(C)} = \dfrac{23/256}{35/256} = \dfrac{23}{35}$

g $P(C \mid P) = \dfrac{P(C \cap P)}{P(P)} = \dfrac{12/256}{101/256} = \dfrac{12}{101}$

h $P(B^C) = 1 - P(B) = 1 - \dfrac{67}{256} = \dfrac{189}{256}$

4.67 Define the following events:
   A: player makes first free throw
   B: player makes second free throw
The probabilities of events A and B will depend on which player is shooting.

a $P(A \cap B) = P(A)P(B) = .85(.85) = .7225$, since the free throws are independent.

b The event that Lamar makes exactly one of the two free throws will occur if he makes the first and misses the second, or vice versa. Then
$$P(\text{makes exactly one}) = P(A \cap B^C) + P(A^C \cap B)$$
$$= .62(.38) + .38(.62) = .4712$$

c This probability is the intersection of the individual probabilities for both Kobe and Lamar.

$P(\text{Kobe makes both and Lamar makes neither}) = [.85(.85)][.38(.38)] = .1043$

4.69 **a** Use the Law of Total Probability, writing
$$P(A) = P(S_1)P(A \mid S_1) + P(S_2)P(A \mid S_2) = .7(.2) + .3(.3) = .23$$

b   Use the results of part a in the form of Bayes' Rule:

$$P(S_i \mid A) = \frac{P(S_i)P(A \mid S_i)}{P(S_1)P(A \mid S_1) + P(S_2)P(A \mid S_2)}$$

For $i = 1$, $P(S_1 \mid A) = \dfrac{.7(.2)}{.7(.2) + .3(.3)} = \dfrac{.14}{.23} = .6087$

For $i = 2$, $P(S_2 \mid A) = \dfrac{.3(.3)}{.7(.2) + .3(.3)} = \dfrac{.09}{.23} = .3913$

4.71   Use the Law of Total Probability, writing $P(A) = P(S_1)P(A \mid S_1) + P(S_2)P(A \mid S_2) = .6(.3) + .4(.5) = .38$.

4.73   Define   A: machine produces a defective item
                          B: worker follows instructions

Then $P(A \mid B) = .01,\ P(B) = .90,\ P(A \mid B^C) = .03,\ P(B^C) = .10$. The probability of interest is

$$P(A) = P(A \cap B) + P(A \cap B^C)$$
$$= P(A \mid B)P(B) + P(A \mid B^C)P(B^C)$$
$$= .01(.90) + .03(.10) = .012$$

4.75   Define   L: play goes to the left
                          R: play goes to the right
                          S: right guard shifts his stance

a   It is given that $P(L) = .3,\ P(R) = .7,\ P(S \mid R) = .8,\ P(S^C \mid L) = .9,\ P(S \mid L) = .1$, and $P(S^C \mid R) = .2$. Using Bayes' Rule,

$$P(L \mid S^C) = \frac{P(L)P(S^C \mid L)}{P(L)P(S^C \mid L) + P(R)P(S^C \mid R)} = \frac{.3(.9)}{.3(.9) + .7(.2)} = \frac{.27}{.41} = .6585$$

b   From part a, $P(R \mid S^C) = 1 - P(L \mid S^C) = 1 - .6585 = .3415$.

c   Given that the guard takes a balanced stance, it is more likely (.6585 versus .3415) that the play will go to the left.

4.77   The probability of interest is $P(A \mid H)$ which can be calculated using Bayes' Rule and the probabilities given in the exercise.

$$P(A \mid H) = \frac{P(A)P(H \mid A)}{P(A)P(H \mid A) + P(B)P(H \mid B) + P(C)P(H \mid C)}$$
$$= \frac{.01(.90)}{.01(.90) + .005(.95) + .02(.75)} = \frac{.009}{.02875} = .3130$$

4.79   **a**   Using the probability table,

$P(D) = .08 + .02 = .10$ $P(D^C) = 1 - P(D) = 1 - .10 = .90$

$P(N \mid D^C) = \dfrac{P(N \cap D^C)}{P(D^C)} = \dfrac{.85}{.90} = .94$ $P(N \mid D) = \dfrac{P(N \cap D)}{P(D)} = \dfrac{.02}{.10} = .20$

b   Using Bayes' Rule, $P(D \mid N) = \dfrac{P(D)P(N \mid D)}{P(D)P(N \mid D) + P(D^C)P(N \mid D^C)} = \dfrac{.10(.20)}{.10(.20) + .90(.94)} = .023$

c   Using the definition of conditional probability,

$$P(D \mid N) = \frac{P(N \cap D)}{P(N)} = \frac{.02}{.87} = .023$$

d $P(\text{false positive}) = P(P \mid D^C) = \dfrac{P(P \cap D^C)}{P(D^C)} = \dfrac{.05}{.90} = .056$

48

e $P(\text{false negative}) = P(N \mid D) = \dfrac{P(N \cap D)}{P(D)} = \dfrac{.02}{.10} = .20$

f   The probability of a false negative is quite high, and would cause concern about the reliability of the screening method.

4.81   **a**   The increase in length of life achieved by a cancer patient as a result of surgery is a continuous random variable, since an increase in life (measured in units of time) can take on any of an infinite number of values in a particular interval.

b   The tensile strength, in pounds per square inch, of one-inch diameter steel wire cable is a continuous random variable.

c   The number of deer killed per year in a state wildlife preserve is a discrete random variable taking the values 0, 1, 2, ...

d   The number of overdue accounts in a department store at a particular point in time is a discrete random variable, taking the values 0, 1, 2, ...

e   Blood pressure is a continuous random variable.

4.83   **a**   Since one of the requirements of a probability distribution is that $\sum_x p(x) = 1$, we need

$$p(3) = 1 - (.1 + .3 + .3 + .1) = 1 - .8 = .2$$

b   The probability histogram is shown below.

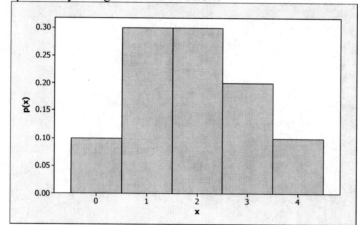

c   For the random variable x given here,
$$\mu = E(x) = \sum xp(x) = 0(.1) + 1(.3) + \cdots + 4(.1) = 1.9$$
The variance of x is defined as
$$\sigma^2 = E\left[(x - \mu)^2\right] = \sum (x - \mu)^2 p(x) = (0 - 1.9)^2(.1) + (1 - 1.9)^2(.3) + \cdots + (4 - 1.9)^2(.1) = 1.29$$
and $\sigma = \sqrt{1.29} = 1.136$.

d   Using the table form of the probability distribution given in the exercise, $P(x > 2) = .2 + .1 = .3$.

e $P(x \le 3) = 1 - P(x = 4) = 1 - .1 = .9$.

4.85   For the probability distribution given in this exercise,
$$\mu = E(x) = \sum xp(x) = 0(.1) + 1(.4) + 2(.4) + 3(.1) = 1.5.$$

4.87   **a**   Define
    T: person admits to texting while driving
    N: person does not admit to texting while driving
There are eight simple events in the experiment:
    TTT   TTN

49

$$\begin{array}{ll} TNN & TNT \\ NTN & NTT \\ NNT & NNN \end{array}$$

and the probabilities for x = number who admit to texting while driving = 0, 1, 2, 3 are shown below.

$$P(x=0) = P(NNN) = (.53)^3 = .148877$$

$$P(x=1) = P(TNN) + P(NTN) + P(NNT) = 3(.47)(.53)^2 = .396069$$

$$P(x=2) = P(NTT) + P(TTN) + P(TNT) = 3(.47)^2(.53) = .351231$$

$$P(x=3) = P(TTT) = (.47)^3 = .103823$$

b   The probability histogram is shown below.

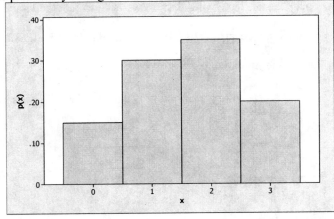

c $P(x=1) = .396069$

d   The average value of x is

$$\mu = E(x) = \sum x p(x) = 0(.148877) + 1(.396069) + 2(.351231) + 3(.103823) = 1.41$$

The variance of x is

$$\sigma^2 = E\left[(x-\mu)^2\right] = \sum (x-\mu)^2 p(x)$$
$$= (0-1.41)^2(.148877) + (1-1.41)^2(.396069) + (2-1.41)^2(.351231) + (3-1.41)^2(.103823)$$
$$= .7473$$

and

$$\sigma = \sqrt{.7473} = .864 \,.$$

4.89   **a-b**   Let W_1 and W_2 be the two women while M_1, M_2 and M_3 are the three men. There are 10 ways to choose the two people to fill the positions. Let x be the number of women chosen. The 10 equally likely simple events are:

E_1: W_1W_2 $(x=2)$	E_6: W_2M_2 $(x=1)$
E_2: W_1M_1 $(x=1)$	E_7: W_2M_3 $(x=1)$
E_3: W_1M_2 $(x=1)$	E_8: M_1M_2 $(x=0)$
E_4: W_1M_3 $(x=1)$	E_9: M_1M_3 $(x=0)$
E_5: W_2M_1 $(x=1)$	E_{10}: M_2M_3 $(x=0)$

The probability distribution for x is then $p(0) = 3/10$, $p(1) = 6/10$, $p(2) = 1/10$. The probability histogram is shown on the next page.

50

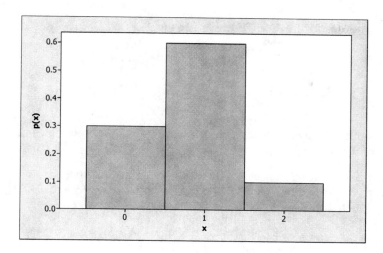

4.91   Let x be the number of drillings until the first success (oil is struck). It is given that the probability of striking oil is $P(O) = .1$, so that the probability of no oil is $P(N) = .9$

a $p(1) = P[\text{oil struck on first drilling}] = P(O) = .1$

   $p(2) = P[\text{oil struck on second drilling}]$. This is the probability that oil is not found on the first drilling, but is found on the second drilling. Using the Multiplication Law,
$$p(2) = P(NO) = (.9)(.1) = .09.$$
Finally, $p(3) = P(NNO) = (.9)(.9)(.1) = .081$.

b-c   For the first success to occur on trial x, $(x-1)$ failures must occur before the first success. Thus,
$$p(x) = P(NNN\ldots NNO) = (.9)^{x-1}(.1)$$
since there are $(x-1)$ N's in the sequence. The probability histogram is shown below.

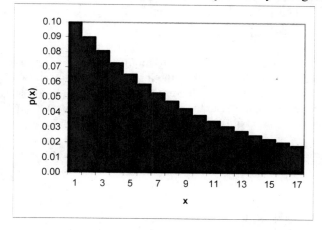

4.93   Refer to Exercise 4.92. For a general value $P(A) = p$ and $P(B) = 1-p$, we showed that the probability distribution for x is
$$p(3) = p^3 + (1-p)^3$$
$$p(4) = 3p^3(1-p) + 3p(1-p)^3$$
$$p(5) = 6p^3(1-p)^2 + 6p^2(1-p)^3$$

For the three values of p given in this exercise, the probability distributions and values of $E(x)$ are given in the table on the next page.

$P(A)=.6$		$P(A)=.5$		$P(A)=.9$	
x	$p(x)$	x	$p(x)$	x	$p(x)$
3	.2800	3	.25	3	.7300
4	.3744	4	.375	4	.2214
5	.3456	5	.375	5	.0486

a $E(x)=4.0656$ **b** $E(x)=4.125$ **c** $E(x)=3.3186$

d Notice that as the probability of winning a single set, *P(A)*, increases, the expected number of sets to win the match, *E(x)*, decreases.

4.95 The random variable G, total gain to the insurance company, will be D if there is no theft, but D – 50,000 if there is a theft during a given year. These two events will occur with probability .99 and .01, respectively. Hence, the probability distribution for G is given below.

G	$p(G)$
D	.99
D – 50,000	.01

The expected gain is
$$E(G)=\sum Gp(G)=.99D+.01(D-50,000)$$
$$= D-50,000$$

In order that $E(G)=1000$, it is necessary to have $1000 = D-500$ or $D=\$1500$.

4.97 **a** Refer to the probability distribution for *x* given in this exercise.
   $P(\text{no coffee breaks}) = P(x=0)=.28$

b $P(\text{more than 2 coffee breaks}) = P(x>2)=.12+.05+.01=.18$

c $\mu = E(x)=\sum xp(x)=0(.28)+1(.37)+\cdots+5(.01)=1.32$

$\sigma^2 = \sum (x-\mu)^2\, p(x)=(0-1.32)^2(.28)+(1-1.32)^2(.37)+\cdots+(5-1.32)^2(.01)=1.4376$ and
$\sigma = \sqrt{1.4376}=1.199$.

d Calculate $\mu \pm 2\sigma = 1.32 \pm 2.398$ or -1.078 to 3.718. Then, referring to the probability distribution of *x*, $P[-1.078 \le x \le 3.718] = P[0 \le x \le 3]=.28+.37+.17+.12=.94$.

4.99 We are asked to find the premium that the insurance company should charge in order to break even. Let *c* be the unknown value of the premium and *x* be the gain to the insurance company caused by marketing the new product. There are three possible values for *x*. If the product is a failure or moderately successful, *x* will be negative; if the product is a success, the insurance company will gain the amount of the premium and *x* will be positive. The probability distribution for *x* follows:

x	$p(x)$
c	.94
$-800,000 + c$	.01
$-250,000 + c$	.05

In order to break even, $E(x)=\sum xp(x)=0$. Therefore,

$$.94(c)+.01(-800,000+c)+(.05)(-250,000+c)=0$$
$$-8000-12,500+(.01+.05+.94)c=0$$
$$c=20,500$$

Hence, the insurance company should charge a premium of $20,500.

4.101 Define the following events:
   A: employee fails to report fraud
   B: employee suffers reprisal
It is given that $P(B\,|\,A^C)=.23$ and $P(A)=.69$. The probability of interest is
$$P(A^C \cap B)=P(B\,|\,A^C)P(A^C)=.23(.31)=.0713$$

52

4.103 Refer to Exercise 4.102. There are four tablets to choose from, call them C_1, C_2, A_1 and A_2. The resulting simple events are then all possible ordered pairs which can be formed from the four choices.

$$(C_1C_2) \quad (C_2C_1) \quad (A_1C_1) \quad (A_2C_1)$$
$$(C_1A_1) \quad (C_2A_1) \quad (A_1C_2) \quad (A_2C_2)$$
$$(C_1A_2) \quad (C_2A_2) \quad (A_1A_2) \quad (A_2A_1)$$

These 12 simple events, each have equal probability 1/12. By summing the probabilities of simple events in the events of interest we have

$$P(A) = 6/12 = 1/2 \qquad\qquad P(A \cup B) = 10/12 = 5/6$$
$$P(B) = 8/12 = 2/3 \qquad\qquad P(C) = 2/12 = 1/6$$
$$P(A \cap B) = 4/12 = 1/3 \qquad\quad P(A \cap C) = 0$$
$$P(A \cup C) = 8/12 = 2/3$$

4.105 Two systems are selected from seven, three of which are defective. Denote the seven systems as G_1, G_2, G_3, G_4, D_1, D_2, D_3 according to whether they are good or defective. Each simple event will represent a particular pair of systems chosen for testing, and the sample space, consisting of 21 pairs, is shown below.

$$G_1G_2 \quad G_1D_1 \quad G_2D_3 \quad G_4D_2 \quad G_1G_3 \quad G_1G_2 \quad G_3D_1 \quad G_4D_3$$
$$G_1G_4 \quad G_1D_3 \quad G_3D_2 \quad D_1D_2 \quad G_2G_3 \quad G_2D_1 \quad G_3D_3 \quad D_1D_3$$
$$G_2G_4 \quad G_2D_2 \quad G_4D_1 \quad D_2D_3 \quad G_3G_4$$

Note that the two systems are drawn simultaneously and that order is unimportant in identifying a simple event. Hence, the pairs G_1G_2 and G_2G_1 are not considered to represent two different simple events. The event A, "no defectives are selected", consists of the simple events G_1G_2, G_1G_3, G_1G_4, G_2G_3, G_2G_4, G_3G_4. Since the systems are selected at random, any pair has an equal probability of being selected. Hence, the probability assigned to each simple event is 1/21 and $P(A) = 6/21 = 2/7$.

4.107 The random variable x, defined as the number of householders insured against fire, can assume the values 0, 1, 2, 3 or 4. The probability that, on any of the four draws, an insured person is found is .6; hence, the probability of finding an uninsured person is .4. Note that each numerical event represents the intersection of the results of four independent draws.

1 $P[x = 0] = (.4)(.4)(.4)(.4) = .0256$, since all four people must be uninsured.

2 $P[x = 1] = 4(.6)(.4)(.4)(.4) = .1536$ (Note: the 4 appears in this expression because $x = 1$ is the union of four mutually exclusive events. These represent the 4 ways to choose the single insured person from the fours.)

3 $P[x = 2] = 6(.6)(.6)(.4)(.4) = .3456$, since the two insured people can be chosen in any of 6 ways.

4 $P[x = 3] = 4(.6)^3(.4) = .3456$ and $P[x = 4] = (.6)^4 = .1296$.

Then

$$P[\text{at least three insured}] = p(3) + p(4) = .3456 + .1296 = .4752$$

4.109 If a $5 bet is placed on the number 18, the gambler will either win $175 $(35 \times \$5)$ with probability 1/38 or lose $5 with probability 37/38. Hence, the probability distribution for x, the gambler's gain is

x	$p(x)$
-5	37/38
175	1/38

The expected gain is $\mu = E(x) = \sum xp(x) = -5(37/38) + 175(1/38) = \$ - 0.26$.

The expected gain is in fact negative, a loss of $0.26.

4.111 Similar to Exercise 4.7. An investor can invest in three of five recommended stocks. Unknown to him, only 2 out of 5 will show substantial profit. Let P_1 and P_2 be the two profitable stocks. A typical simple event might be $(P_1P_2N_3)$, which represents the selection of two profitable and one nonprofitable stock. The ten simple events are listed below:

53

$$\begin{array}{llll} E_1:\ (P_1P_2N_1) & E_2:\ (P_1P_2N_2) & E_3:\ (P_1P_2N_3) & E_4:\ (P_2N_1N_2) \\ E_5:\ (P_2N_1N_3) & E_6:\ (P_2N_2N_3) & E_7:\ (N_1N_2N_3) & E_8:\ (P_1N_1N_2) \\ E_9:\ (P_1N_1N_3) & E_{10}:\ (P_1N_2N_3) \end{array}$$

Then $P[\text{investor selects two profitable stocks}] = P(E_1) + P(E_2) + P(E_3) = 3/10$ since the simple events are equally likely, with $P(E_i) = 1/10$. Similarly,

$$P[\text{investor selects only one of the profitable stocks}] = P(E_4) + P(E_5) + P(E_6) + P(E_8) + P(E_9) + P(E_{10}) = 6/10$$

4.113 **a** Define the following events:

B_1: client buys on first contact
B_2: client buys on second contact

Since the client may buy on either the first of the second contact, the desired probability is
$P[\text{client will buy}] = P[\text{client buys on first contact}] + P[\text{client doesn't buy on first, but buys on second}]$

$$= P(B_1) + (1 - P(B_1))P(B_2) = .4 + (1-.4)(.55)$$
$$= .73$$

b The probability that the client will not buy is one minus the probability that the client will buy, or $1 - .73 = .27$.

4.115 Define the following events: A: first system fails
    B: second system fails

A and B are independent and $P(A) = P(B) = .001$. To determine the probability that the combined missile system does not fail, we use the complement of this event; that is,
$$P[\text{system does not fail}] = 1 - P[\text{system fails}] = 1 - P(A \cap B)$$
$$= 1 - P(A)P(B) = 1 - (.001)^2$$
$$= .999999$$

4.117 Each ball can be chosen from the set (4, 6) and there are three such balls. Hence, there are a total of $2(2)(2) = 8$ potential winning numbers.

4.119 **a** Use the Law of Total Probability, with $P(B) = .47$, $P(F) = .53$, $P(MCL\,|\,B) = .39$, $P(ACL\,|\,B) = .61$, $P(MCL\,|\,F) = .33$, and $P(ACL\,|\,F) = .67$. Then
$$P(MCL) = P(MCL \cap B) + P(MCL \cap F) = P(B)P(MCL\,|\,B) + P(F)P(MCL\,|\,F)$$
$$= .47(.39) + .53(.33) = .3582$$

b Use Bayes' Rule, $P(F\,|\,MCL) = \dfrac{P(MCL \cap F)}{P(MCL)} = \dfrac{P(F)P(MCL\,|\,F)}{P(MCL)} = \dfrac{.53(.33)}{.3582} = .4883$.

c Use Bayes' Rule, $P(B\,|\,ACL) = \dfrac{P(B)P(ACL\,|\,B)}{P(B)P(ACL\,|\,B) + P(F)P(ACL\,|\,F)} = \dfrac{.47(.61)}{.47(.61) + .53(.67)} = .4467$

4.121 **a** Consider a single trial which consists of tossing two coins. A match occurs when either HH or TT is observed. Hence, the probability of a match on a single trial is $P(HH) + P(TT) = 1/4 + 1/4 = 1/2$. Let MMM denote the event "match on trials 1, 2, and 3".

Then $P(MMM) = P(M)P(M)P(M) = (1/2)^3 = 1/8$.

b On a single trial the event A, "two tails are observed" has probability $P(A) = P(TT) = 1/4$. Hence, in three trials

$$P(AAA) = P(A)P(A)P(A) = (1/4)^3 = 1/64$$

c This low probability would not suggest collusion, since the probability of three matches is low only if we assume that each student is merely guessing at each answer. If the students have studied together or if they both know the correct answer, the probability of a match on a single trial is no longer 1/2, but is substantially higher. Hence, the occurrence of three matches is not unusual.

4.123 Define   R: the employee remains 10 years or more

    a The probability that the man will stay less than 10 years is $P(R^C) = 1 - P(R) = 1 - 1/6 = 5/6$

    b The probability that the man and the woman, acting independently, will both work less than 10 years

    is $$P(R^C R^C) = P(R^C) P(R^C) = (5/6)^2 = 25/36$$

    c The probability that either or both people work 10 years or more is

$$1 - P(R^C R^C) = 1 - (5/6)^2 = 1 - 25/36 = 11/36$$

4.125 Define the events: A: the man waits five minutes or longer
                               B: the woman waits five minutes or longer

The two events are independent, and $P(A) = P(B) = .2$.

    a $P(A^C) = 1 - P(A) = .8$

    b $P(A^C B^C) = P(A^C) P(B^C) = (.8)(.8) = .64$

    c $P[\text{at least one waits five minutes or longer}]$

$$= 1 - P[\text{neither waits five minutes or longer}] = 1 - P(A^C B^C) = 1 - .64 = .36$$

4.127 It is given that 40% of all people in a community favor the development of a mass transit system. Thus, given a person selected at random, the probability that the person will favor the system is .4. Since the pollings are independent events, when four people are selected at random,

$$P[\text{all 4 favor the system}] = (.4)^4 = .0256$$

Similarly, $P[\text{none favor the system}] = (1 - .4)^4 = .1296$

4.129 Since the first pooled test is positive, we are interested in the probability of requiring five single tests to detect the disease in the single affected person. There are (5)(4)(3)(2)(1) ways of ordering the five tests, and there are 4(3)(2)(1) ways of ordering the tests so that the diseased person is given the final test. Hence, the desired probability is $\frac{4!}{5!} = \frac{1}{5} = .2$.

If two people are diseased, six tests are needed if the last two tests are given to the diseased people. There are 3(2)(1) ways of ordering the tests of the other three people and 2(1) ways of ordering the tests of the two diseased people. Hence, the probability that six tests will be needed is $\frac{2!3!}{5!} = \frac{1}{10} = .1$.

4.131 The necessary probabilities can be found by summing the necessary cells in the probability table and dividing by 220, the total number of firms.

    a $P(R) = \frac{114}{220} = .5182$ **b** $P(T \cap F) = \frac{25}{220} = .1136$

    c $P(F^C) = \frac{156}{220} = .7091$ **d** $P(T \mid F) = \frac{P(T \cap F)}{P(F)} = \frac{25/220}{64/220} = \frac{25}{64} = .3906$

4.133 **a** Define P: shopper prefers Pepsi and C: shopper prefers Coke. Then if there is actually no difference in the taste, $P(P) = P(C) = 1/2$ and

$$P(\text{all four prefer Pepsi}) = P(PPPP) = [P(P)]^4 = \left(\frac{1}{2}\right)^4 = \frac{1}{16} = .0625$$

b

$$P(\text{exactly one prefers Pepsi}) = P(PCCC) + P(CPCC) + P(CCPC) + P(CCCP)$$

$$= 4P(P)[P(C)]^3 = 4\left(\frac{1}{2}\right)\left(\frac{1}{2}\right)^3 = \frac{4}{16} = .25$$

4.135 **a** There are six volunteers, from whom we must choose two people for the committee. The number of choices is then

$$C_2^6 = \frac{6(5)}{2(1)} = 15$$

If the number of women chosen from the two women is x, and the number of men chosen from the four men must be $2 - x$. Then

$$P(x = 0) = \frac{C_0^2 C_2^4}{15} = \frac{6}{15}$$

$$P(x = 1) = \frac{C_1^2 C_1^4}{15} = \frac{8}{15}$$

$$P(x = 2) = \frac{C_2^2 C_0^4}{15} = \frac{1}{15}$$

and the probability distribution for x is shown in the table.

x	0	1	2
$p(x)$	6/15	8/15	1/15

b The probability that both orchestra representatives will be women is $P(x = 2) = 1/15$.

c $\mu = E(x) = \sum x p(x) = 0\left(\frac{6}{15}\right) + 1\left(\frac{8}{15}\right) + 2\left(\frac{1}{15}\right) = \frac{10}{15} = \frac{2}{3}$

$$\sigma^2 = \sum (x - \mu)^2 p(x) = (0 - \frac{2}{3})^2 \left(\frac{6}{15}\right) + (1 - \frac{2}{3})^2 \left(\frac{8}{15}\right) + (2 - \frac{2}{3})^2 \left(\frac{1}{15}\right) = \frac{48}{135} = \frac{16}{45}$$

4.137 The necessary probabilities can be found by summing the necessary cells in the probability table and dividing by 200, the total number of individuals.

a $P(Y) = \dfrac{96}{200} = .48$

b $P(M \cap AE) = \dfrac{20}{200} = .10$

c $P(M \mid S) = \dfrac{P(M \cap S)}{P(S)} = \dfrac{16/200}{61/200} = \dfrac{16}{61} = .262$

5: Several Useful Discrete Distributions

5.1 These probabilities can be found individually using the binomial formula, or alternatively using the cumulative binomial tables (Table 1 in Appendix I). The values in Table 1 for $n = 8$ and $p = .7$ are shown in the table below.

k	0	1	2	3	4	5	6	7	8
$P(x \leq k)$	.000	.001	.011	.058	.194	.448	.745	.942	1.000

a Using the formula,
$$P(x \leq 3) = C_0^8(.7)^0(.3)^8 + C_1^8(.7)^1(.3)^7 + C_2^8(.7)^2(.3)^6 + C_3^8(.7)^3(.3)^5 = .058$$
or read directly from the table as $P(x \leq 3) = .058$.

b Using the formula,
$$P(x \geq 3) = \sum_{x=3}^{8} C_x^8(.7)^x(.3)^{8-x} = .989$$

or use Table 1, writing $P(x \geq 3) = 1 - P(x \leq 2) = 1 - .011 = .989$

e Using the formula,
$$P(x < 3) = C_0^8(.7)^0(.3)^8 + C_1^8(.7)^1(.3)^7 + C_2^8(.7)^2(.3)^6 = .011$$
or use Table 1, writing $P(x < 3) = P(x \leq 2) = .011$.

d Using the formula, $P(x = 3) = C_3^8(.7)^3(.3)^5 = 56(.343)(.00243) = .047$ or use Table 1, writing $P(x = 3) = P(x \leq 3) - P(x \leq 2) = .058 - .011 = .047$.

f Using the formula,
$$P(3 \leq x \leq 5) = C_3^8(.7)^3(.3)^5 + C_4^8(.7)^4(.3)^3 + C_5^8(.7)^5(.3)^3 = .437$$

or use Table 1, writing $P(3 \leq x \leq 5) = P(x \leq 5) - P(x \leq 2) = .448 - .011 = .437$.

5.3 **a** $C_2^8(.3)^2(.7)^6 = \dfrac{8(7)}{2(1)}(.09)(.117649) = .2965$

b $C_0^4(.05)^0(.95)^4 = (.95)^4 = .8145$

c $C_3^{10}(.5)^3(.5)^7 = \dfrac{10(9)(8)}{3(2)(1)}(.5)^{10} = .1172$

d $C_1^7(.2)^1(.8)^6 = 7(.2)(.8)^6 = .3670$

5.5 **a** For $n = 7$ and $p = .3$, $P(x = 4) = C_4^7(.3)^4(.7)^3 = .097$.

b These probabilities can be found individually using the binomial formula, or alternatively using the cumulative binomial tables in Appendix I.
$$P(x \leq 1) = p(0) + p(1)$$
$$= C_0^7(.3)^0(.7)^7 + C_1^7(.3)^1(.7)^6$$
$$= (.7)^7 + 7(.3)(.7)^6 = .08235 + .24706 = .329$$
or directly from the binomial tables in the row marked $a = 1$.

c Refer to part **b**. $P(x > 1) = 1 - P(x \leq 1) = 1 - .329 = .671$.

d $\mu = np = 7(.3) = 2.1$

e $\sigma = \sqrt{npq} = \sqrt{7(.3)(.7)} = \sqrt{1.47} = 1.212$

5.7   Notice that when $p = .8$, $p(x) = C_x^6(.8)^x(.2)^{6-x}$. In Exercise 5.6, with $p = .2$, $p(x) = C_x^6(.2)^x(.8)^{6-x}$. The probability that $x = k$ when $p = .8$ --- $C_k^6(.8)^k(.2)^{n-k}$ --- is the same as the probability that $x = n - k$ when $p = .2$ --- $C_{n-k}^6(.2)^{n-k}(.8)^k$. This follows because

$$C_k^n = \frac{n!}{k!(n-k)!} = C_{n-k}^n$$

Therefore, the probabilities $p(x)$ for a binomial random variable x when $n = 6$ and $p = .8$ will be the mirror images of those found in Exercise 5.6 as shown in the table. The probability histogram is shown on the next page.

x	0	1	2	3	4	5	6
$p(x)$	.000	.002	.015	.082	.246	.393	.262

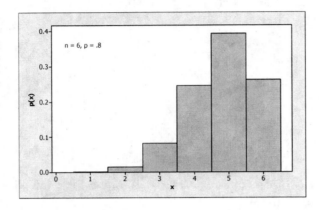

5.9   **a**   For $n = 10$ and $p = .4$, $P(x = 4) = C_4^{10}(.4)^4(.6)^6 = .251$.

   b   To calculate $P(x \geq 4) = p(4) + p(5) + \cdots + p(10)$ it is easiest to write
$$P(x \geq 4) = 1 - P(x < 4) = 1 - P(x \leq 3).$$

These probabilities can be found individually using the binomial formula, or alternatively using the cumulative binomial tables in Appendix I.

$$P(x = 0) = C_0^{10}(.4)^0(.6)^{10} = .006 \qquad P(x = 1) = C_1^{10}(.4)^1(.6)^9 = .040$$

$$P(x = 2) = C_2^{10}(.4)^2(.6)^8 = .121 \qquad P(x = 3) = C_3^{10}(.4)^3(.6)^7 = .215$$

The sum of these probabilities gives $P(x \leq 3) = .382$ and $P(x \geq 4) = 1 - .382 = .618$.

   c   Use the results of parts **a** and **b**.
$$P(x > 4) = 1 - P(x \leq 4) = 1 - (.382 + .251) = .367$$

   d   From part **c**, $P(x \leq 4) = P(x \leq 3) + P(x = 4) = .382 + .251 = .633$.

   e $\mu = np = 10(.4) = 4$

   f $\sigma = \sqrt{npq} = \sqrt{10(.4)(.6)} = \sqrt{2.4} = 1.549$

5.11   **a** $P[x \geq 4] = 1 - P[x \leq 3] = 1 - .099 = .901$

   b $P[x = 2] = P[x \leq 2] - P[x \leq 1] = .017 - .002 = .015$

   c $P[x < 2] = P[x \leq 1] = .002$

   d $P[x > 1] = 1 - P[x \leq 1] = 1 - .002 = .998$

5.13   **a** $P[x < 12] = P[x \leq 11] = .748$

   b $P[x \leq 6] = .610$

58

c $P[x > 4] = 1 - P[x \le 4] = 1 - .633 = .367$

d $P[x \ge 6] = 1 - P[x \le 5] = 1 - .034 = .966$

e $P[3 < x < 7] = P[x \le 6] - P[x \le 3] = .828 - .172 = .656$

5.15　**a**　$\mu = 100(.01) = 1; \ \sigma = \sqrt{100(.01)(.99)} = .99$

　　　　b　$\mu = 100(.9) = 90; \ \sigma = \sqrt{100(.9)(.1)} = 3$

　　　　c　$\mu = 100(.3) = 30; \ \sigma = \sqrt{100(.3)(.7)} = 4.58$

　　　　d　$\mu = 100(.7) = 70; \ \sigma = \sqrt{100(.7)(.3)} = 4.58$

　　　　e　$\mu = 100(.5) = 50; \ \sigma = \sqrt{100(.5)(.5)} = 5$

5.17　**a**　$p(0) = C_0^{20}(.1)^0 (.9)^{20} = .1215767$　　　$p(3) = C_3^{20}(.1)^3 (.9)^{17} = .1901199$

　　　　　$p(1) = C_1^{20}(.1)^1 (.9)^{19} = .2701703$　　　$p(4) = C_4^{20}(.1)^4 (.9)^{16} = .0897788$

　　　　　$p(2) = C_2^{20}(.1)^2 (.9)^{18} = .2851798$

　　so that　$P[x \le 4] = p(0) + p(1) + p(2) + p(3) + p(4) = .9568255$

　　　　b　Using Table 1, Appendix I, $P[x \le 4]$ is read directly as .957.

　　　　c　Adding the entries for $x = 0, 1, 2, 3, 4$, we have $P[x \le 4] = .9569$.

　　　　d　$\mu = np = 20(.1) = 2$ and $\sigma = \sqrt{npq} = \sqrt{1.8} = 1.3416$

　　　　e　For $k = 1$, $\mu \pm \sigma = 2 \pm 1.342$ or .658 to 3.342 so that
$$P[.658 \le x \le 3.342] = P[1 \le x \le 3] = .2702 + .2852 + .1901 = .7455$$
　　　　　For $k = 2$, $\mu \pm 2\sigma = 2 \pm 2.683$ or $-.683$ to 4.683 so that
$$P[-.683 \le x \le 4.683] = P[0 \le x \le 4] = .9569$$
　　　　　For $k = 3$, $\mu \pm 3\sigma = 2 \pm 4.025$ or -2.025 to 6.025 so that
$$P[-2.025 \le x \le 6.025] = P[0 \le x \le 6] = .9977$$

　　　　f　The results are consistent with Tchebysheff's Theorem and the Empirical Rule.

5.19　If the sampling in Exercise 5.18 is conducted with replacement, then x is a binomial random variable with $n = 2$ independent trials, and $p = P[\text{red ball}] = 3/5$, which remains constant from trial to trial.

5.21　Although there are trials (telephone calls) which result in either a person who will answer (S) or a person who will not (F), the number of trials, n, is not fixed in advance. Instead of recording x, the number of *successes* in n trials, you record x, the number of *trials* until the first success. This is *not* a binomial experiment.

5.23　Define x to be the number of alarm systems that are triggered. Then $p = P[\text{alarm is triggered}] = .99$ and $n = 9$. Since there is a table available in Appendix I for $n = 9$ and $p = .99$, you should use it rather than the binomial formula to calculate the necessary probabilities.

　　　　a　$P[\text{at least one alarm is triggered}] = P(x \ge 1) = 1 - P(x = 0) = 1 - .000 = 1.000$.

　　　　b　$P[\text{more than seven}] = P(x > 7) = 1 - P(x \le 7) = 1 - .003 = .997$

　　　　c　$P[\text{eight or fewer}] = P(x \le 8) = .086$

5.25 Define x to be the number of cars that are black. Then $p = P[\text{black}] = .1$ and $n = 25$. Use Table 1 in Appendix I.

    a $P(x \geq 5) = 1 - P(x \leq 4) = 1 - .902 = .098$

    b $P(x \leq 6) = .991$

    c $P(x > 4) = 1 - P(x \leq 4) = 1 - .902 = .098$

    d $P(x = 4) = P(x \leq 4) - P(x \leq 3) = .902 - .764 = .138$

    e $P(3 \leq x \leq 5) = P(x \leq 5) - P(x \leq 2) = .967 - .537 = .430$

    f $P(\text{more than 20 } not \text{ black}) = P(\text{less than 5 black}) = P(x \leq 4) = .902$

5.27 Define a success to be a patient who fails to pay his bill and is eventually forgiven. Assuming that the trials are independent and that p is constant from trial to trial, this problem satisfies the requirements for the binomial experiment with $n = 4$ and $p = .3$. You can use either the binomial formula or Table 1.

    a $P[x = 4] = p(4) = C_4^4 (.3)^4 (.7)^0 = (.3)^4 = .0081$

    b $P[x = 1] = p(1) = C_1^4 (.3)^1 (.7)^3 = 4(.3)(.7)^3 = .4116$

    c $P[x = 0] = C_0^4 (.3)^0 (.7)^4 = (.7)^4 = .2401$

5.29 Define x to be the number of fields infested with whitefly. Then $p = P[\text{infected field}] = .1$ and $n = 100$.

    a $\mu = np = 100(.1) = 10$

    b   Since n is large, this binomial distribution should be fairly mound-shaped, even though $p = .1$. Hence you would expect approximately 95% of the measurements to lie within two standard deviation of the mean with $\sigma = \sqrt{npq} = \sqrt{100(.1)(.9)} = 3$. The limits are calculated as

$$\mu \pm 2\sigma \Rightarrow 10 \pm 6 \text{ or from 4 to 16}$$

    c   From part **b**, a value of $x = 25$ would be very unlikely, assuming that the characteristics of the binomial experiment are met and that $p = .1$. If this value were actually observed, it might be possible that the trials (fields) are not independent. This could easily be the case, since an infestation in one field might quickly spread to a neighboring field. This is evidence of *contagion*.

5.31 Define x to be the number of adults who indicate that back pain was a limiting factor in their athletic activities. Then, $n = 8$ and $p = .6$.

    a $P(x = 8) = C_8^8 (.6)^8 (.4)^0 = .016796$

    b   From the Minitab **Probability Density Function**, read $P(x = 8) = .016796$.

    c   Use the Minitab **Cumulative Distribution Function** to find $P(x \leq 7) = .98320$.

5.33 Define x to be the number of Americans who are "tasters". Then, $n = 20$ and $p = .7$. Using the binomial tables in Appendix I,

    a $P(x \geq 17) = 1 - P(x \leq 16) = 1 - .893 = .107$

    b $P(x \leq 15) = .762$

5.35 Use the Poisson formula with $\mu = 2.5$.

    a $P(x = 0) = \dfrac{2.5^0 e^{-2.5}}{0!} = .082085$ **c** $P(x = 2) = \dfrac{2.5^2 e^{-2.5}}{2!} = .256516$

    b $P(x = 1) = \dfrac{2.5^1 e^{-2.5}}{1!} = .205212$ **d** $P(x \leq 2) = .082085 + .205212 + .256516 = .543813$

5.37   These probabilities should be found using the cumulative Poisson tables (Table 2 in Appendix I). The values in Table 2 for $\mu = 3$ are shown in the table below.

k	0	1	2	3	4	5	6	7	8	9	10
$P(x \leq k)$	.055	.199	.423	.647	.815	.916	.966	.988	.996	.999	1.000

a $P(x \leq 3) = .647$, directly from the table.

b $P(x > 3) = 1 - P(x \leq 3) = 1 - .647 = .353$.

c $P(x = 3) = P(x \leq 3) - P(x \leq 2) = .647 - .423 = .224$.

d $P(3 \leq x \leq 5) = P(x \leq 5) - P(x \leq 2) = .916 - .423 = .493$.

5.39   Using $p(x) = \dfrac{\mu^x e^{-\mu}}{x!} = \dfrac{2^x e^{-2}}{x!}$,

a $P[x = 0] = \dfrac{2^0 e^{-2}}{0!} = .135335$

b $P[x = 1] = \dfrac{2^1 e^{-2}}{1!} = .27067$

c $P[x > 1] = 1 - P[x \leq 1] = 1 - .135335 - .27067 = .593994$

d $P[x = 5] = \dfrac{2^5 e^{-2}}{5!} = .036089$

5.41   **a**   Using Table 1, Appendix I, $P[x \leq 2] = .677$

b   With $\mu = np = 20(.1) = 2$, the approximation is $p(x) \approx \dfrac{2^x e^{-2}}{x!}$. Then

$$P[x \leq 2] \approx \frac{2^0 e^{-0}}{0!} + \frac{2^1 e^{-1}}{1!} + \frac{2^2 e^{-2}}{2!}$$
$$= .135335 + .27067 + .27067 = .676675$$

c   The approximation is quite accurate.

5.43   Let x be the number of misses during a given month. Then x has a Poisson distribution with $\mu = 5$.

a $p(0) = e^{-5} = .0067$ **b** $p(5) = \dfrac{5^5 e^{-5}}{5!} = .1755$

c $P[x \geq 5] = 1 - P[x \leq 4] = 1 - .440 = .560$ from Table 2.

5.45   Let x be the number of injuries per year, with $\mu = 2$.

a $P[x = 2] = P[x \leq 2] - P[x \leq 1] = .677 - .406 = .271$

b $P[x \geq 2] = 1 - P[x \leq 1] = 1 - .406 = .594$

c $P[x \leq 1] = .406$

5.47   The random variable x, number of bacteria, has a Poisson distribution with $\mu = 2$. The probability of interest is

$$P[x \text{ exceeds maximum count}] = P[x > 5]$$

Using the fact that $\mu = 2$ and $\sigma = 1.414$, most of the observations should fall within $\mu \pm 2\sigma$ or 0 to 4. Hence, it is unlikely that x will exceed 5. In fact, the exact Poisson probability is $P[x > 5] = .017$.

5.49   **a** $\quad \dfrac{C_1^2 C_1^1}{C_2^3} = \dfrac{2(1)}{3} = \dfrac{2}{3}$ **b** $\quad \dfrac{C_0^4 C_2^2}{C_2^6} = \dfrac{1(1)}{15} = \dfrac{1}{15}$ **c** $\quad \dfrac{C_2^2 C_1^2}{C_3^4} = \dfrac{1(2)}{4} = \dfrac{1}{2}$

5.51   **a** $\quad \dfrac{C_1^3 C_1^2}{C_2^5} = \dfrac{3(2)}{10} = .6$ **b** $\quad \dfrac{C_2^4 C_1^3}{C_3^7} = \dfrac{6(3)}{35} = .5143$ **c** $\quad \dfrac{C_4^5 C_0^3}{C_4^8} = \dfrac{5(1)}{70} = .0714$

5.53   The formula for $p(x)$ is $p(x) = \dfrac{C_x^4 C_{3-x}^{11}}{C_3^{15}}$ for $x = 0, 1, 2, 3$

     a $\quad p(0) = \dfrac{C_0^4 C_3^{11}}{C_3^{15}} = \dfrac{165}{455} = .36$ $p(1) = \dfrac{C_1^4 C_2^{11}}{C_3^{15}} = \dfrac{220}{455} = .48$

        $p(2) = \dfrac{C_2^4 C_1^{11}}{C_3^{15}} = \dfrac{66}{455} = .15$ $p(3) = \dfrac{C_3^4 C_0^{11}}{C_3^{15}} = \dfrac{4}{455} = .01$

     b   The probability histogram is shown below.

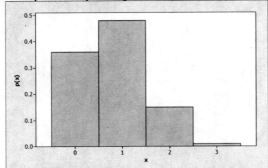

     c   Using the formulas given in Section 5.4.

$$\mu = E(x) = n\left(\frac{M}{N}\right) = 3\left(\frac{4}{15}\right) = .8$$

$$\sigma^2 = n\left(\frac{M}{N}\right)\left(\frac{N-M}{N}\right)\left(\frac{N-n}{N-1}\right) = 3\left(\frac{4}{15}\right)\left(\frac{15-4}{15}\right)\left(\frac{15-3}{15-1}\right) = .50286$$

     d   Calculate the intervals

$$\mu \pm 2\sigma = .8 \pm 2\sqrt{.50286} = .8 \pm 1.418 \text{ or } -.618 \text{ to } 2.218$$

$$\mu \pm 3\sigma = .8 \pm 3\sqrt{.50286} = .8 \pm 1.418 \text{ or } -1.327 \text{ to } 2.927$$

Then,

$$P\left[-.618 \le x \le 2.218\right] = p(0) + p(1) + p(2) = .99$$

$$P\left[-1.327 \le x \le 2.927\right] = p(0) + p(1) + p(2) = .99$$

These results agree with Tchebysheff's Theorem.

5.55   The formula for $p(x)$ is $p(x) = \dfrac{C_x^2 C_{3-x}^4}{C_3^6}$ for $x = $ number of defectives $= 0, 1, 2.$ Then

$$p(0) = \frac{C_0^2 C_3^4}{C_3^6} = \frac{4}{20} = .2 \qquad p(1) = \frac{C_1^2 C_2^4}{C_3^6} = \frac{12}{20} = .6 \qquad p(2) = \frac{C_2^2 C_1^4}{C_3^6} = \frac{4}{20} = .2$$

These results agree with the probabilities calculated in Exercise 4.90.

5.57   **a**   The random variable x has a hypergeometric distribution with $N = 8, M = 5$ and $n = 3.$ Then

$$p(x) = \frac{C_x^5 C_{3-x}^3}{C_3^8} \text{ for } x = 0, 1, 2, 3$$

     b $P(x = 3) = \dfrac{C_3^5 C_0^3}{C_3^8} = \dfrac{10}{56} = .1786$

c $P(x=0) = \dfrac{C_0^5 C_3^3}{C_3^8} = \dfrac{1}{56} = .01786$

d $P(x \le 1) = \dfrac{C_0^5 C_3^3}{C_3^8} + \dfrac{C_1^5 C_2^3}{C_3^8} = \dfrac{1+15}{56} = .2857$

5.59 See Section 5.2 in the text.

5.61 The hypergeometric distribution is appropriate when sampling from a finite rather than an infinite population of successes and failures. In this case, the probability of p of a success is not constant from trial to trial and the binomial distribution is not appropriate.

5.62 The random variable x is defined to be the number of heads observed when a coin is flipped three times. Then $p = P[\text{success}] = P[\text{head}] = 1/2$, $q = 1 - p = 1/2$ and $n = 3$. The binomial formula yields the following results.

a $P[x=0] = p(0) = C_0^3 (1/2)^0 (1/2)^3 = 1/8$ $P[x=1] = p(1) = C_1^3 (1/2)^1 (1/2)^2 = 3/8$

  $P[x=2] = p(2) = C_2^3 (1/2)^2 (1/2)^1 = 3/8$ $P[x=3] = p(3) = C_3^3 (1/2)^3 (1/2)^0 = 1/8$

b The associated probability histogram is shown below.

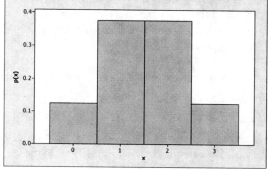

c $\mu = np = 3(1/2) = 1.5$ and $\sigma = \sqrt{npq} = \sqrt{3(1/2)(1/2)} = .866$

d The desired intervals are

  $\mu \pm \sigma = 1.5 \pm .866$ or $.634$ to 2.366

  $\mu \pm 2\sigma = 8 \pm 1.732$ or $-.232$ to 3.232

The values of x which fall in this first interval are $x = 1$ and $x = 2$, and the fraction of measurement in this interval will be $3/8 + 3/8 = 3/4$. The second interval encloses all four values of x and thus the fraction of measurements within 2 standard deviations of the mean will be 1, or 100%. These results are consistent with both Tchebysheff's Theorem and the Empirical Rule.

5.63 The random variable x is defined to be the number of heads observed when a coin is flipped three times. Then $p = P[\text{success}] = P[\text{head}] = .1$, $q = 1 - p = .9$ and $n = 3$. The binomial formula yields the following results.

a $P[x=0] = p(0) = C_0^3 (.1)^0 (.9)^3 = .729$ $P[x=1] = p(1) = C_1^3 (.1)^1 (.9)^2 = .243$

  $P[x=2] = p(2) = C_2^3 (.1)^2 (.9)^1 = .027$ $P[x=3] = p(3) = C_3^3 (.1)^3 (.9)^0 = .001$

b   Note that the probability distribution is no longer symmetric as it was in Exercise 5.62; that is, since the probability of observing a head is so small, the probability of observing a small number of heads on three flips is increased (see the figure below).

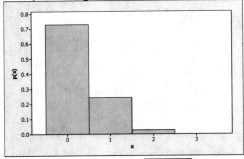

c $\mu = np = 3(.1) = .3$ and $\sigma = \sqrt{npq} = \sqrt{3(.1)(.9)} = .520$

d   The desired intervals are

$$\mu \pm \sigma = .3 \pm .520 \quad \text{or} \quad -.220 \text{ to } .820$$

$$\mu \pm 2\sigma = .3 \pm 1.04 \quad \text{or} \quad -.740 \text{ to } 1.34$$

The only value of x which falls in this first interval is $x = 0$, and the fraction of measurements in this interval will be .729. The values of $x = 0$ and $x = 1$ are enclosed by the second interval, so that $.729 + .243 = .972$ of the measurements fall within two standard deviations of the mean, consistent with both Tchebysheff's Theorem and the Empirical Rule.

5.65   Define x to be the number supporting the commissioner's claim, with $p = .8$ and $n = 25$.

a   Using the binomial tables for $n = 25$, $P[x \geq 22] = 1 - P[x \leq 21] = 1 - .766 = .234$

b $P[x = 22] = P[x \leq 22] - P[x \leq 21] = .902 - .766 = .136$

c   The probability of observing an event as extreme as $x = 22$ (or more extreme) is quite high assuming that $p = .8$. Hence, this is not an unlikely event and we would not doubt the claim.

5.67   From Exercise 5.66, the probability that a person chooses a 4, 5 or 6 is 3/10, assuming that the 10 digits are all equally likely to be chosen. Define x to be the number of people who choose a 4, 5 or 6 in the sample of $n = 20$. Then x has a binomial distribution with $p = .3$.

a $P[x \geq 8] = 1 - P[x \leq 7] = 1 - .772 = .228$

b   Observing eight or more people choosing a 4,5 or 6 is not an unlikely event, assuming that the integers are all equally likely. Therefore, there is no evidence to indicate that people are more likely to choose these three numbers than any others.

5.69   It is given that x = number of patients with a psychosomatic problem, $n = 25$, and $p = P[\text{patient has psychosomatic problem}]$. A psychiatrist wishes to determine whether or not $p = .8$.

a   Assuming that the psychiatrist is correct (that is, $p = .8$), the expected value of x is

$E(x) = np = 25(.8) = 20$.

b $\sigma^2 = npq = 25(.8)(.2) = 4$

c   Given that $p = .8$, $P[x \leq 14] = .006$ from Table 1 in Appendix I.

d   Assuming that the psychiatrist is correct, the probability of observing $x = 14$ or the more unlikely values, $x = 0, 1, 2, \ldots, 13$ is very unlikely. Hence, one of two conclusions can be drawn. Either we have observed a very unlikely event, or the psychiatrist is incorrect and p is actually less than .8. We would probably conclude that the psychiatrist is incorrect. The probability that we have made an incorrect decision is

$$P[x \leq 14 \text{ given } p = .8] = .006$$

which is quite small.

5.71   Define x to be the number of students 30 years or older, with $n = 200$ and $p = P[\text{student is 30+ years}] = .25$.

   a   Since x has a binomial distribution, $\mu = np = 200(.25) = 50$ and $\sigma = \sqrt{npq} = \sqrt{200(.25)(.75)} = 6.124$.

   b   The observed value, $x = 35$, lies

$$\frac{35 - 50}{6.124} = -2.45$$

standard deviations below the mean. It is unlikely that $p = .25$.

5.73   **a**   If there is no preference for either design, then $p = P[\text{choose the second design}] = .5$.

   b   Using the results of part **a** and $n = 25, \mu = np = 25(.5) = 12.5$ and $\sigma = \sqrt{npq} = \sqrt{6.25} = 2.5$.

   c   The observed value, $x = 20$, lies

$$z = \frac{20 - 12.5}{2.5} = 3$$

standard deviations above the mean. This is an unlikely event, assuming that $p = .5$. We would probably conclude that there is a preference for the second design and that $p > .5$.

5.75   **a**   The random variable x, the number of plants with red petals, has a binomial distribution with $n = 10$ and $p = P[\text{red petals}] = .75$.

   b   Since the value $p = .75$ is not given in Table 1, you must use the binomial formula to calculate

$$P(x \geq 9) = C_9^{10}(.75)^9(.25)^1 + C_{10}^{10}(.75)^{10}(.25)^0 = .1877 + .0563 = .2440$$

   c $P(x \leq 1) = C_0^{10}(.75)^0(.25)^{10} + C_1^{10}(.75)^1(.25)^9 = .0000296$.

   d   Refer to part **c**. The probability of observing $x = 1$ or something even more unlikely $(x = 0)$ is very small – $.0000296$. This is a highly unlikely event if in fact $p = .75$. Perhaps there has been a nonrandom choice of seeds, or the 75% figure is not correct for this particular genetic cross.

5.77   Let x be the number of lost calls in a series of $n = 11$ trials. If the coin is fair, then $p = \frac{1}{2}$ and x has a binomial distribution.

   a $P[x = 11] = \left(\frac{1}{2}\right)^{11} = \frac{1}{2048}$ which is the same as odds of 1:2047.

   b   If $n = 13$, $P[x = 13] = \left(\frac{1}{2}\right)^{13} = \frac{1}{8192}$ which is a very unlikely event.

5.79   **a**   The distribution of x is actually hypergeometric, with $N = 50$, $n = 10$ and $M =$ number of phones not working (due to defective internal components) in the carton. That is,

$$p(x) = \frac{C_x^M C_{10-x}^{50-M}}{C_{10}^{50}} \text{ for } x = 0, 1, 2, ..., M$$

   b   The carton will only be shipped if there are no defective phones in the sample of $n = 10$. If $M = 2$,

$$P(x = 0) = \frac{C_0^2 C_{10}^{48}}{C_{10}^{50}} = \frac{48!40!}{38!50!} = \frac{40(39)}{50(49)} = .6367$$

   c   If there are 4 defectives in the carton, then $M = 4$ and the probability that the carton is shipped is

$$P(x = 0) = \frac{C_0^6 C_{10}^{46}}{C_{10}^{50}} = \frac{46!40!}{36!50!} = \frac{40(39)(38)(37)}{50(49)(48)(47)} = .3968$$

65

If there are 6 defectives, then $M = 6$ and the probability that the carton is shipped is

$$P(x=0) = \frac{C_0^6 C_{10}^{44}}{C_{10}^{50}} = \frac{44!40!}{34!50!} = \frac{40(39)(38)(37)(36)(35)}{50(49)(48)(47)(46)(45)} = .2415$$

5.81 The random variable x, the number of offspring with Tay-Sachs disease, has a binomial distribution with $n = 3$ and $p = .25$. Use the binomial formula.

a $P(x=3) = C_3^3 (.25)^3 (.75)^0 = (.25)^3 = .015625$

b $P(x=1) = C_1^3 (.25)^1 (.75)^2 = 3(.25)(.75)^2 = .421875$

c Remember that the trials are independent. Hence, the occurrence of Tay-Sachs in the first two children has no effect on the third child, and $P(\text{third child develops Tay-Sachs}) = .25$.

5.83 **a** The random variable x, the number of tasters who pick the correct sample, has a binomial distribution with $n=5$ and, if there is no difference in the taste of the three samples,

$$p = P(\text{taster picks the correct sample}) = \frac{1}{3}$$

b The probability that exactly one of the five tasters chooses the latest batch as different from the others is

$$P(x=1) = C_1^5 \left(\frac{1}{3}\right)^1 \left(\frac{2}{3}\right)^4 = .3292$$

c The probability that at least one of the tasters chooses the latest batch as different from the others is

$$P(x \geq 1) = 1 - P(x=0) = 1 - C_0^5 \left(\frac{1}{3}\right)^0 \left(\frac{2}{3}\right)^5 = .8683$$

5.85 Refer to Exercise 5.84. The random variable x, the number of questionnaires that are filled out and returned, has a binomial distribution with $n = 20$ and $p = .7$.

a The average value of x is $\mu = np = 20(.7) = 14$.

b The standard deviation of x is $\sigma = \sqrt{npq} = \sqrt{20(.7)(.3)} = \sqrt{4.2} = 2.049$.

c The z-score corresponding to $x = 10$ is

$$z = \frac{x - \mu}{\sigma} = \frac{10 - 14}{2.049} = -1.95.$$

Since this z-score does not exceed 3 in absolute value, we would not consider the value $x = 10$ to be an unusual observation.

5.87 The random variable x has a Poisson distribution with $\mu = 2$. Use Table 2 in Appendix I or the Poisson formula to find the following probabilities.

a $P(x=0) = \frac{2^0 e^{-2}}{0!} = e^{-2} = .135335$

b $P(x \leq 2) = \frac{2^0 e^{-2}}{0!} + \frac{2^1 e^{-2}}{1!} + \frac{2^2 e^{-2}}{2!}$

$$= .135335 + .270671 + .270671 = .676676$$

5.89 The random variable x, the number of subjects who revert to their first learned method under stress, has a binomial distribution with $n = 6$ and $p = .8$. The probability that at least five of the six subjects revert to their first learned method is

$$P(x \geq 5) = 1 - P(x \leq 4) = 1 - .345 = .655$$

66

5.91   The random variable x, the number of California homeowners with earthquake insurance, has a binomial distribution with $n = 15$ and $p = .1$.

a $P(x \geq 1) = 1 - P(x = 0) = 1 - .206 = .794$

b $P(x \geq 4) = 1 - P(x \leq 3) = 1 - .944 = .056$

c   Calculate $\mu = np = 15(.1) = 1.5$ and $\sigma = \sqrt{npq} = \sqrt{15(.1)(.9)} = 1.1619$. Then approximately 95% of the values of x should lie in the interval

$$\mu \pm 2\sigma \Rightarrow 1.5 \pm 2(1.1619) \Rightarrow -.82 \text{ to } 3.82.$$

or between 0 and 3.

5.93   The random variable x, the number of women who get fast food when they are too busy, has a binomial distribution with $n = 100$ and $p = .36$.

a   The average value of x is $\mu = np = 100(.36) = 36$.

b   The standard deviation of x is $\sigma = \sqrt{npq} = \sqrt{100(.36)(.64)} = \sqrt{23.04} = 4.8$.

c   The z-score corresponding to $x = 49$ is

$$z = \frac{x - \mu}{\sigma} = \frac{49 - 36}{4.8} = 2.71$$

Since this value is not greater than 3 in absolute value, it is not an extremely unusual observation. It is between 2 and 3, however, so it is somewhat unusual.

5.95   The random variable x, the number of consumers who say they are committed to living with fewer credit cards, has a binomial distribution with $n = 400$ and $p = .60$.

a   The average value of x is $\mu = np = 400(.60) = 240$.

b   The standard deviation of x is $\sigma = \sqrt{npq} = \sqrt{400(.60)(.40)} = \sqrt{96} = 9.798$.

c   Most values of x should lie within two standard deviations of the mean, or

$$\mu \pm 2\sigma \Rightarrow 240 \pm 2(9.798) \Rightarrow 240 \pm 19.596 \text{ or } 220.404 \text{ to } 259.596.$$

Since the binomial random variable can only take on integer values, the range of values is 221 to 259.

d   The z-score corresponding to $x = 200$ is

$$z = \frac{x - \mu}{\sigma} = \frac{200 - 240}{9.798} = -4.08$$

Since this value is greater than 3 in absolute value, it is an unlikely observation. Perhaps the 60% figure quoted in the exercise is too high.

5.97   In part c, Exercise 5.96, we found that the probability of less than two successful operations is $P[x < 2] = p(0) + p(1) = C_0^5 (.8)^0 (.2)^5 + C_1^5 (.8)^1 (.2)^4 = .00032 + .0064 = .0067$. This is a very rare event, and the fact that it has occurred leads to one of two conclusions. Either the success rate is in fact 80% and a very rare occurrence has been observed, or the success rate is less than 80% and we are observing a likely event under the latter assumption. The second conclusion is more likely. We would probably conclude that the success rate for this team is less than 80% and would put little faith in the team.

5.99   Define x to be the number of young adults who prefer McDonald's. Then x has a binomial distribution with $n = 10$ and $p = .5$.

a $P(x > 6) = 1 - P(x \leq 6) = 1 - .828 = .172$.

b $P(4 \leq x \leq 6) = P(x \leq 6) - P(x \leq 3) = .828 - .172 = .656$.

c   If 4 prefer Burger King, then 6 prefer McDonalds, and vice versa. The probability is the same as that calculated in part **b**, since $p = .5$.

6: The Normal Probability Distribution

6.1   The first few exercises are designed to provide practice for the student in evaluating areas under the normal curve. The following notes may be of some assistance.

1   Table 3, Appendix I tabulates the cumulative area under a standard normal curve to the left of a specified value of z.

2   Since the total area under the curve is one, the total area lying to the right of a specified value of z and the total area to its left must add to 1. Thus, in order to calculate a "tail area", such as the one shown in Figure 6.1, the value of $z = z_0$ will be indexed in Table 3, and the area that is obtained will be subtracted from 1. Denote the area obtained by indexing $z = z_0$ in Table 3 by $A(z_0)$ and the desired area by A. Then, in the above example, $A = 1 - A(z_0)$.

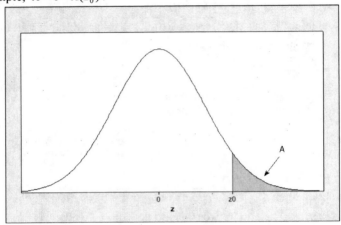

3   To find the area under the standard normal curve between two values, z_1 and z_2, calculate the difference in their cumulative areas, $A = A(z_2) - A(z_1)$.

4   Note that z, similar to x, is actually a random variable which may take on an infinite number of values, both positive and negative. Negative values of z lie to the left of the mean, $z = 0$, and positive values lie to the right.

a $P(z < 2) = P(z < 2.00) = A(2.00) = .9772$

b $P(z > 1.16) = 1 - P(z \le 1.16) = 1 - A(1.16) = 1 - .8770 = .1230$

c $P(-2.33 < z < 2.33) = A(2.33) - A(-2.33) = .9901 - .0099 = .9802$

d $P(z < 1.88) = A(1.88) = .9699$

6.3   **a**   It is necessary to find the area to the left of $z = 1.6$. That is, $A = A(1.6) = .9452$.

b   The area to the left of $z = 1.83$ is $A = A(1.83) = .9664$.

c $A = A(.90) = .8159$

d $A = A(4.58) \approx 1$. Notice that the values in Table 3 approach 1 as the value of z increases. When the value of z is larger than $z = 3.49$ (the largest value in the table), we can assume that the area to its left is approximately 1.

6.5   **a** $P(-1.43 < z < .68) = A(.68) - A(-1.43) = .7517 - .0764 = .6753$

b $P(.58 < z < 1.74) = A(1.74) - A(.58) = .9591 - .7190 = .2401$

c $P(-1.55 < z < -.44) = A(-.44) - A(-1.55) = .3300 - .0606 = .2694$

d $P(z > 1.34) = 1 - A(1.34) = 1 - .9099 = .0901$

68

e Since the value of $z = -4.32$ is not recorded in Table 3, you can assume that the area to the left of $z = -4.32$ is very close to 0. Then

$$P(z < -4.32) \approx 0$$

6.7 Now we are asked to find the z-value corresponding to a particular area.

a We need to find a z_0 such that $P(z > z_0) = .025$. This is equivalent to finding an indexed area of $1 - .025 = .975$. Search the interior of Table 3 until you find the four-digit number **.9750**. The corresponding z-value is **1.96**; that is, $A(1.96) = .9750$. Therefore, $z_0 = 1.96$ is the desired z-value (see the figure below).

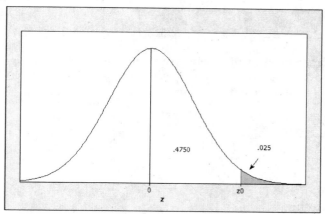

b We need to find a z_0 such that $P(z < z_0) = .9251$ (see below). Using Table 3, we find a value such that the indexed area is .9251. The corresponding z-value is $z_0 = 1.44$.

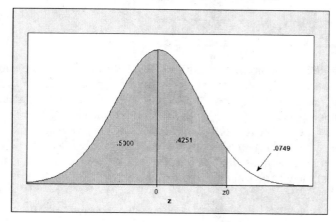

6.9 **a** Similar to Exercise 6.7(b). The value of z_0 must be positive and $A(z_0) = .9505$. Hence, $z_0 = 1.65$.

b It is given that the area to the left of z_0 is .05, shown as A_1 in the figure that follows. The desired value is not tabulated in Table 3 but lies halfway between two tabulated values, .0505 and .0495. Hence, using linear interpolation (as we did in Exercise 6.6(b)) we choose a value of z_0 that lies halfway between -1.64 and -1.65, or $z_0 = -1.645$.

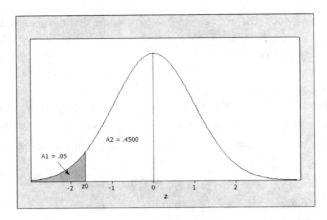

6.11 The pth percentile of the standard normal distribution is a value of z which has area $p/100$ to its left. Since all four percentiles in this exercise are greater than the 50th percentile, the value of z will all lie to the right of $z = 0$, as shown for the 90th percentile in the figure below.

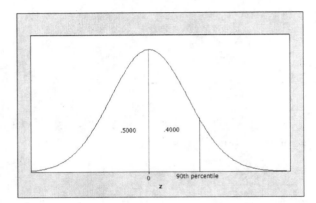

a From the figure, the area to the left of the 90th percentile is .9000. From Table 3, the appropriate value of z is closest to $z = 1.28$ with area .8997. Hence the 90th percentile is approximately $z = 1.28$.

b As in part **a**, the area to the left of the 95th percentile is .9500. From Table 3, the appropriate value of z is found using linear interpolation (see Exercise 6.9(b)) as $z = 1.645$. Hence the 95th percentile is $z = 1.645$.

c The area to the left of the 98th percentile is .9800. From Table 3, the appropriate value of z is closest to $z = 2.05$ with area .9798. Hence the 98th percentile is approximately $z = 2.05$.

d The area to the left of the 99th percentile is .9900. From Table 3, the appropriate value of z is closest to $z = 2.33$ with area .9901. Hence the 99th percentile is approximately $z = 2.33$.

6.13 Since $z = (x - \mu)/\sigma$ measures the number of standard deviations an observation lies from its mean, it can be used to standardize any normal random variable x so that Table 3 can be used.

a Calculate $z_1 = \dfrac{1.00 - 1.20}{.15} = -1.33$ and $z_2 = \dfrac{1.10 - 1.20}{.15} = -.67$. Then

$$P(1.00 < x < 1.10) = P(-1.33 < z < -.67) = .2514 - .0918 = .1596$$

b Calculate $z = \dfrac{x - \mu}{\sigma} = \dfrac{1.38 - 1.20}{.15} = 1.2$. Then

$$P(x > 1.38) = P(z > 1.2) = 1 - .8849 = .1151$$

c Calculate $z_1 = \dfrac{1.35 - 1.20}{.15} = 1$ and $z_2 = \dfrac{1.50 - 1.20}{.15} = 2$. Then

$$P(1.35 < x < 1.50) = P(1 < z < 2) = .9772 - .8413 = .1359$$

70

6.15   The 99th percentile of the standard normal distribution was found in Exercise 6.11(d) to be $z = 2.33$. Since the relationship between the general normal random variable x and the standard normal z is $z = \dfrac{x - \mu}{\sigma}$, the corresponding percentile for this general normal random variable is found by solving for $x = \mu + z\sigma$;

$$2.33 = \frac{x - 35}{10}$$
$$x - 35 = 23.3 \quad \text{or} \quad x = 58.3$$

6.17   The random variable x is normal with unknown μ and σ. However, it is given that

$$P(x > 4) = P\left(z > \frac{4 - \mu}{\sigma}\right) = .9772 \quad \text{and} \quad P(x > 5) = P\left(z > \frac{5 - \mu}{\sigma}\right) = .9332 .$$ These probabilities are shown in the figure below.

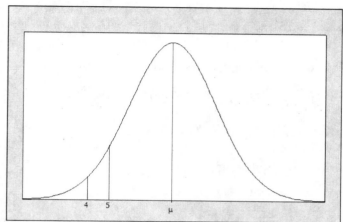

The value $\dfrac{4 - \mu}{\sigma}$ is negative, with $A\left(\dfrac{4 - \mu}{\sigma}\right) = 1 - .9772 = .0228$ or $\dfrac{4 - \mu}{\sigma} = -2$ (i)

The value $\dfrac{5 - \mu}{\sigma}$ is also negative, with $A\left(\dfrac{5 - \mu}{\sigma}\right) = 1 - .9332 = .0668$ or $\dfrac{5 - \mu}{\sigma} = -1.5$ (ii)

Equations (i) and (ii) provide two equations in two unknowns which can be solved simultaneously for μ and σ. From (i), $\sigma = \dfrac{\mu - 4}{2}$ which, when substituted into (ii) yields

$$5 - \mu = -1.5\left(\frac{\mu - 4}{2}\right)$$
$$10 - 2\mu = -1.5\mu + 6$$
$$\mu = 8$$

and from (i), $\sigma = \dfrac{8 - 4}{2} = 2$.

6.19   The random variable x, the height of a male human, has a normal distribution with $\mu = 69.5$ and $\sigma = 3.5$.

a   A height of 6'0" represents $6(12) = 72$ inches, so that

$$P(x > 72) = P\left(z > \frac{72 - 69.5}{3.5}\right) = P(z > .71) = 1 - .7611 = .2389$$

b   Heights of 5'8" and 6'1" represent $5(12) + 8 = 68$ and $6(12) + 1 = 73$ inches, respectively. Then

71

$$P(68 < x < 73) = P\left(\frac{68-69.5}{3.5} < z < \frac{73-69.5}{3.5}\right) = P(-.43 < z < 1.00) = .8413 - .3336 = .5077$$

c A height of 6'1" represents $6(12) + 1 = 73$ inches, which has a z-value of

$$z = \frac{73-69.5}{3.5} = 1.00$$

This would not be considered an unusually large value, since it is less than two standard deviations from the mean.

d The probability that a man is 6'0" or taller was found in part **a** to be .2389, which is not an unusual occurrence. However, if you define y to be the number of men in a random sample of size $n = 43$ who are 6'0" or taller, then y has a binomial distribution with mean $\mu = np = 43(.2389) = 10.2727$ and standard deviation $\sigma = \sqrt{npq} = \sqrt{43(.2389)(.7611)} = 2.796$. The value $y = 18$ lies

$$\frac{y-\mu}{\sigma} = \frac{18-10.2727}{2.796} = 2.76$$

standard deviations from the mean, and would be considered a somewhat unusual occurrence for the general population of male humans. Perhaps our presidents do not represent a *random* sample from this population.

6.21 The random variable x, cerebral blood flow, has a normal distribution with $\mu = 74$ and $\sigma = 16$.

a $P(60 < x < 80) = P\left(\frac{60-74}{16} < z < \frac{80-74}{16}\right) = P(-.88 < z < .38) = .6480 - .1894 = .4586$

b $P(x > 100) = P\left(z > \frac{100-74}{16}\right) = P(z > 1.62) = 1 - .9474 = .0526$

c $P(x < 40) = P\left(z < \frac{40-74}{16}\right) = P(z < -2.12) = .0170$

6.23 The random variable x, total weight of 8 people, has a mean of $\mu = 1200$ and a standard deviation $\sigma = 99$. It is necessary to find $P(x > 1300)$ and $P(x > 1500)$ if the distribution of x is approximately normal. Refer to the figure below.

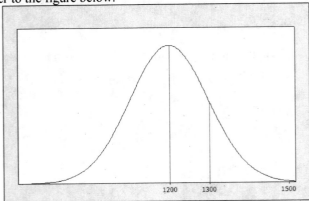

The z-value corresponding to $x_1 = 1300$ is $z_1 = \frac{x_1 - \mu}{\sigma} = \frac{1300 - 1200}{99} = 1.01$. Hence,

$$P(x > 1300) = P(z > 1.01) = 1 - A(1.01) = 1 - .8438 = .1562.$$

Similarly, the z-value corresponding to $x_2 = 1500$ is $z_2 = \frac{x_2 - \mu}{\sigma} = \frac{1500 - 1200}{99} = 3.03$.

and $P(x > 1500) = P(z > 3.03) = 1 - A(3.03) = 1 - .9988 = .0012.$

72

6.25   It is given that x, the unsupported stem diameter of a sunflower plant, is normally distributed with $\mu = 35$ and $\sigma = 3$.

a $P(x > 40) = P\left(z > \dfrac{40 - 35}{3}\right) = P(z > 1.67) = 1 - .9525 = .0475$

b   From part **a**, the probability that one plant has stem diameter of more than 40 mm is .0475. Since the two plants are independent, the probability that two plants both have diameters of more than 40 mm is
$$(.0475)(.0475) = .00226$$

c   Since 95% of all measurements for a normal random variable lie within 1.96 standard deviations of the mean, the necessary interval is
$$\mu \pm 1.96\sigma \quad \Rightarrow \quad 35 \pm 1.96(3) \quad \Rightarrow \quad 35 \pm 5.88$$
or in the interval 29.12 to 40.88.

d   The 90th percentile of the standard normal distribution was found in Exercise 6.11(a) to be $z = 1.28$. Since the relationship between the general normal random variable x and the standard normal z is $z = \dfrac{x - \mu}{\sigma}$, the corresponding percentile for this general normal random variable is found by solving for $x = \mu + z\sigma$.

$$x = 35 + 1.28(3) \quad \text{or} \quad x = 38.84$$

6.27   **a**   It is given that the prime interest rate forecasts, x, are approximately normal with mean $\mu = 4.75$ and standard deviation $\sigma = 0.2$. It is necessary to determine the probability that x exceeds 4.25. Calculate $z = \dfrac{x - \mu}{\sigma} = \dfrac{4.25 - 4.75}{0.2} = -2.5$. Then
$$P(x > 4.25) = P(z > -2.5) = 1 - .0062 = .9938 \, .$$

b   Calculate $z = \dfrac{x - \mu}{\sigma} = \dfrac{4.375 - 4.75}{0.2} = -1.88$. Then
$$P(x < 4.375) = P(z < -1.88) = .0301 \, .$$

6.29   It is given that the counts of the number of bacteria are normally distributed with $\mu = 85$ and $\sigma = 9$. The z-value corresponding to $x = 100$ is $z = \dfrac{x - \mu}{\sigma} = \dfrac{100 - 85}{9} = 1.67$ and
$$P(x > 100) = P(z > 1.67) = 1 - .9525 = .0475$$

6.31   Let w be the number of words specified in the contract. Then x, the number of words in the manuscript, is normally distributed with $\mu = w + 20{,}000$ and $\sigma = 10{,}000$. The publisher would like to specify w so that
$$P(x < 100{,}000) = .95 \, .$$

As in Exercise 6.30, calculate
$$z = \frac{100{,}0000 - (w + 20{,}000)}{10{,}000} = \frac{80{,}000 - w}{10{,}000} \, .$$

Then $P(x < 100{,}000) = P\left(z < \dfrac{80{,}000 - w}{10{,}000}\right) = .95$. It is necessary that $z_0 = (80{,}000 - w)/10{,}000$ be such that
$$P(z < z_0) = .95 \quad \Rightarrow \quad A(z_0) = .9500 \quad \text{or} \quad z_0 = 1.645 \, .$$

Hence,
$$\frac{80{,}000 - w}{10{,}000} = 1.645 \quad \text{or} \quad w = 63{,}550 \, .$$

6.33 The amount of money spent between 4 and 6 pm on Sundays is normally distributed with $\mu = 85$ and $\sigma = 20$.

a The z-value corresponding to $x = 95$ is $z = \dfrac{x-\mu}{\sigma} = \dfrac{95-85}{20} = 0.5$. Then

$$P(x > 95) = P(z > 0.5) = 1 - .6915 = .3085$$

b The z-value corresponding to $x = 115$ is $z = \dfrac{x-\mu}{\sigma} = \dfrac{115-85}{20} = 1.5$. Then

$$P(95 < x < 115) = P(0.5 < z < 1.5) = .9332 - .6915 = .2417$$

c First, find $P(x > 115) = P(z > 1.5) = 1 - .9332 = .0668$ for a single shopper. For two shoppers, use the Multiplication Rule.

$P(\text{both shoppers spend more than \$115}) = P(\text{1st spends more than \$115}) \times P(\text{2nd spends more than \$115})$
$$= (.0668)(.0668) = .0045$$

6.35 **a** Consider a binomial random variable with $n = 25$ and $p = .6$. To determine if the normal approximation is appropriate, calculate $np = 15$ and $nq = 10$. Since both np and nq are greater than 5, the normal approximation is appropriate.

b Calculate $\mu = np = 15$ and $\sigma = \sqrt{npq} = \sqrt{25(.6)(.4)} = 2.449$

c To find the probability of more than 9 successes, we need to include the values $x = 10, 11, \ldots 25$. To include the entire block of probability for the first value of $x = 10$, we need to start at 9.5. Then the probability of more than 9 successes is approximated as

$$P(x > 9.5) = P\left(z > \dfrac{9.5-15}{2.449}\right) = P(z > -2.25) = 1 - .0122 = .9878.$$

6.37 **a** The normal approximation will be appropriate if both np and nq are greater than 5. For this binomial experiment,

$$np = 25(.3) = 7.5 \quad \text{and} \quad nq = 25(.7) = 17.5$$

and the normal approximation is appropriate.

b For the binomial random variable, $\mu = np = 7.5$ and $\sigma = \sqrt{npq} = \sqrt{25(.3)(.7)} = 2.291$.

c The probability of interest is the area under the binomial probability histogram corresponding to the rectangles $x = 6, 7, 8$ and 9 in the figure below.

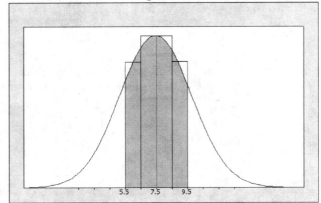

To approximate this area, use the "correction for continuity" and find the area under a normal curve with mean $\mu = 7.5$ and $\sigma = 2.291$ between $x_1 = 5.5$ and $x_2 = 9.5$. The z-values corresponding to the two values of x are

$$z_1 = \dfrac{5.5-7.5}{2.291} = -.87 \quad \text{and} \quad z_2 = \dfrac{9.5-7.5}{2.291} = .87$$

The approximating probability is $P(5.5 < x < 9.5) = P(-.87 < z < .87) = .8078 - .1922 = .6156$.

74

d From Table 1, Appendix I,

$$P(6 \le x \le 9) = P(x \le 9) - P(x \le 5) = .811 - .193 = .618$$

which is not too far from the approximate probability calculated in part **c**.

6.39 Similar to Exercise 6.37.

a The approximating probability will be $P(x > 22.5)$ where x has a normal distribution with $\mu = 100(.2) = 20$ and $\sigma = \sqrt{100(.2)(.8)} = 4$. Then

$$P(x > 22.5) = P\left(z > \frac{22.5 - 20}{4}\right) = P(z > .62) = 1 - .7324 = .2676$$

b The approximating probability is now $P(x > 21.5)$ since the entire rectangle corresponding to $x = 22$ must be included.

$$P(x > 21.5) = P\left(z > \frac{21.5 - 20}{4}\right) = P(z > .38) = 1 - .6480 = .3520$$

c To include the entire rectangles for $x = 21$ and $x = 24$, the approximating probability is

$$P(20.5 < x < 24.5) = P(.12 < z < 1.12) = .8686 - .5478 = .3208$$

d To include the entire rectangle for $x = 25$, the approximating probability is

$$P(x < 25.5) = P(z < 1.38) = .9162$$

6.41 Using the binomial tables for $n = 20$ and $p = .3$, you can verify that

a $P(x = 5) = P(x \le 5) - P(x \le 4) = .416 - .238 = .178$

b $P(x \ge 7) = 1 - P(x \le 6) = 1 - .608 = .392$

6.43 Similar to previous exercises.

a With $n = 20$ and $p = .4$, $P(x \ge 10) = 1 - P(x \le 9) = 1 - .755 = .245$.

b To use the normal approximation, find the mean and standard deviation of this binomial random variable:

$$\mu = np = 20(.4) = 8 \text{ and } \sigma = \sqrt{npq} = \sqrt{20(.4)(.6)} = \sqrt{4.2} = 2.191.$$

Using the continuity correction, it is necessary to find the area to the right of 9.5. The z-value corresponding to $x = 9.5$ is $\qquad z = \dfrac{9.5 - 8}{2.191} = .68$ and

$$P(x \ge 10) \approx P(z > .68) = 1 - .7517 = .2483.$$

Note that the normal approximation is very close to the exact binomial probability.

6.45 **a** The approximating probability will be $P(x < 29.5)$ where x has a normal distribution with $\mu = 50(.78) = 39$ and $\sigma = \sqrt{50(.78)(.22)} = 2.929$. Then

$$P(x < 29.5) = P\left(z < \frac{29.5 - 39}{2.929}\right) = P(z < -3.24) = .0006$$

b The approximating probability is

$$P(x > 40.5) = P\left(z > \frac{40.5 - 39}{2.929}\right) = P(z > .51) = 1 - .6950 = .3050$$

c If more than 10 individuals *do not* believe that recycling does not make the biggest difference, then less than $50 - 10 = 40$ *do* believe it. The approximating probability is

$$P(x < 39.5) = P\left(z < \frac{39.5 - 39}{2.929}\right) = P(z < 0.17) = .5675$$

6.47   Define x to be the number of guests claiming a reservation at the motel. Then
$p = P[\text{guest claims reservation}] = 1 - .1 = .9$ and $n = 215$. The motel has only 200 rooms. Hence, if
$x > 200$, a guest will not receive a room. The probability of interest is then $P(x \le 200)$. Using the
normal approximation, calculate
$$\mu = np = 215(.9) = 193.5 \text{ and } \sigma = \sqrt{215(.9)(.1)} = \sqrt{19.35} = 4.399$$
The probability $P(x \le 200)$ is approximated by the area under the appropriate normal curve to the left of
200.5. The z-value corresponding to $x = 200.5$ is $z = \dfrac{200.5 - 193.5}{\sqrt{19.35}} = 1.59$ and
$$P(x \le 200) \approx P(z < 1.59) = .9441$$

6.49   Define x to be the number of elections in which the taller candidate won. If Americans are not biased by
height, then the random variable x has a binomial distribution with $n = 49$ and $p = .5$. Calculate
$$\mu = np = 49(.5) = 24.5 \text{ and } \sigma = \sqrt{49(.5)(.5)} = \sqrt{12.25} = 3.5$$
a   Using the normal approximation with correction for continuity, we find the area to the right of
$x = 25.5$:
$$P(x > 25.5) = P\left(z > \frac{25.5 - 24.5}{3.5}\right) = P(z > 0.29) = 1 - .6141 = .3859$$
b   Since the occurrence of 26 out of 49 taller choices is not unusual, based on the results of part **a**, it
appears that Americans do not consider height when casting a vote for a candidate.

6.51   Define x to be the number of consumers who preferred a *Pepsi* product. Then the random variable x has a
binomial distribution with $n = 500$ and $p = .25$, if *Pepsi's* market share is indeed 25%. Calculate
$$\mu = np = 500(.25) = 125 \text{ and } \sigma = \sqrt{500(.25)(.75)} = \sqrt{93.75} = 9.682$$
a   Using the normal approximation with correction for continuity, we find the area between $x = 149.5$ and
$x = 150.5$:
$$P(149.5 < x < 150.5) = P\left(\frac{149.5 - 125}{9.682} < z < \frac{150.5 - 125}{9.682}\right) = P(2.53 < z < 2.63) = .9957 - .9943 = .0014$$
b   Find the area between $x = 119.5$ and $x = 150.5$:
$$P(119.5 < x < 150.5) = P\left(\frac{119.5 - 125}{9.682} < z < \frac{150.5 - 125}{9.682}\right) = P(-.57 < z < 2.63) = .9957 - .2843 = .7114$$
c   Find the area to the left of $x = 149.5$:
$$P(x < 149.5) = P\left(z < \frac{149.5 - 125}{9.682}\right) = P(z < 2.53) = .9943$$
d   The value $x = 232$ lies $z = \dfrac{232 - 125}{9.682} = 11.05$ standard deviations above the mean, if *Pepsi's* market
share is indeed 25%. This is such an unusual occurrence that we would conclude that *Pepsi's* market share
is higher than claimed.

6.53   Refer to Exercise 6.52, and let x be the number of working women who put in more than 40 hours per week
on the job. Then x has a binomial distribution with $n = 50$ and $p = .62$.
a   The average value of x is $\mu = np = 50(.62) = 31$.
b   The standard deviation of x is $\sigma = \sqrt{npq} = \sqrt{50(.62)(.38)} = 3.432$.
c   The z-score for $x = 25$ is $z = \dfrac{x - \mu}{\sigma} = \dfrac{25 - 31}{3.432} = -1.75$ which is within two standard deviations of the
mean. This is not considered an unusual occurrence.

6.55   Use Table 3, subtracting successive entries.

    a $P(-2.0 < z < 2.0) = .9772 - .0228 = .9544$

    b $P(-2.3 < z < -1.5) = .0668 - .0107 = .0561$

6.57   **a** It is given that $P(z > z_0) = .9750$, so that $A(z_0) = P(z < z_0) = 1 - .9750 = .0250$. Looking in the interior of Table 3 for .0250, we find that $A(-1.96) = .0250$. That is, $z_0 = -1.96$.

    b It is given that $P(z > z_0) = .3594$, so that $A(z_0) = P(z < z_0) = 1 - .3594 = .6406$. Looking in the interior of Table 3 for .6406, we find that $A(.36) = .6406$. That is, $z_0 = .36$.

6.59   It is given that x is approximately normally distributed with $\mu = 5$ and $\sigma = 2$.

    a Calculate $z = \dfrac{x - \mu}{\sigma} = \dfrac{1.2 - 5}{2} = -1.9$ and $z = \dfrac{x - \mu}{\sigma} = \dfrac{10 - 5}{2} = 2.5$.

    Then $P(1.2 < x < 10) = P(-1.9 < z < 2.5) = .9938 - .0287 = .9651$

    b Calculate $z = \dfrac{x - \mu}{\sigma} = \dfrac{7.5 - 5}{2} = 1.25$.

    Then $P(x > 7.5) = P(z > 1.25) = 1 - A(1.25) = 1 - .8944 = .1056$

    c Calculate $z = \dfrac{x - \mu}{\sigma} = \dfrac{0 - 5}{2} = -2.5$

    Then $P(x \le 0) = P(z < -2.5) = A(-2.5) = .0062$

6.61   **a** The area to the left of $z = 1.2$ is $A(1.2) = .8849$.

    b The area to the left of $z = -.9$ is $A(-.9) = .1841$.

    c $A(1.46) = .9279$

    d $A(-.42) = .3372$

6.63   **a** $P(z \ge -.75) = 1 - .2266 = .7734$

    b $P(z < 1.35) = .9115$

6.65 $P(-1.48 \le z \le 1.48) = .9306 - .0694 = .8612$

6.67   It is given that x is approximately normally distributed with $\mu = 75$ and $\sigma = 12$.

    a Calculate $z = \dfrac{x - \mu}{\sigma} = \dfrac{60 - 75}{12} = -1.25$.

    Then $P(x < 60) = P(z < -1.25) = .1056$

    b $P(x > 60) = 1 - P(x < 60) = 1 - .1056 = .8944$

    c If the bit is replaced after more than 90 hours, then $x > 90$. Calculate $z = \dfrac{x - \mu}{\sigma} = \dfrac{90 - 75}{12} = 1.25$.

    Then $P(x > 90) = P(z > 1.25) = 1 - .8944 = .1056$

6.69   For this exercise, it is given that the population of bolt diameters is normally distributed with $\mu = .498$ and $\sigma = .002$. Thus, no correction for continuity is necessary. The fraction of acceptable bolts will be those which lie in the interval from .496 to .504. All others are unacceptable. The desired fraction of acceptable bolts is calculated, and the fraction of unacceptable bolts (shaded in the figure on the next page) is obtained by subtracting from the total probability, which is 1.

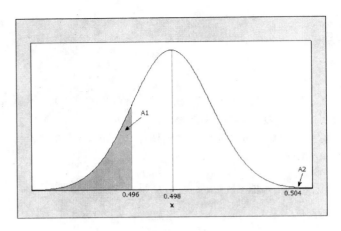

The fraction of acceptable bolts is then

$$P(.496 \le x \le .504) = P\left(\frac{.496 - .498}{.002} \le z \le \frac{.504 - .498}{.002}\right)$$

$$= P(-1 \le z \le 3) = .9987 - .1587 = .8400$$

and the fraction of unacceptable bolts is $1 - .84 = .16$.

6.71 The random variable x is approximately normally distributed with $\mu = 1230$ and $\sigma = 120$.

a The z-value corresponding to $x = 1400$ is $\quad z = \frac{x - \mu}{\sigma} = \frac{1400 - 1230}{120} = 1.42$ and

$$P(x > 1400) = P(z > 1.42) = 1 - .9222 = .0778.$$

b The z-value corresponding to $x = 1000$ is $\quad z = \frac{x - \mu}{\sigma} = \frac{1000 - 1230}{120} = -1.92$ and

$$P[\text{restaurant does not break even}] = P(x < 300) = P(z < -1.92) = .0274.$$

6.73 It is given that x is normally distributed with $\mu = 30$ and $\sigma = 11$. The probability of interest is

$$P(x > 50) = P\left(z > \frac{50 - 30}{11}\right) = P(z > 1.82) = 1 - .9656 = .0344$$

6.75 It is given that $\mu = 1.4$ and $\sigma = .7$. If we assume that x, the service time for one vehicle, is normally distributed, the probability of interest is

$$P(x > 1.6) = P\left(z > \frac{1.6 - 1.4}{.7}\right) = P(z > .29) = 1 - .6141 = .3859$$

6.77 **a** No. The errors in forecasting would tend to be skewed, with more underestimates than overestimates.
b Let x be the number of estimates that are in error by more than 15%. Then x has a binomial distribution with $n = 100$ and $p = P[\text{estimate is in error by more than 15%}] = .5$. Using the normal approximation to the binomial distribution,

$$P(x > 60) \approx P\left(z > \frac{60.5 - 100(.5)}{\sqrt{100(.5)(.5)}}\right) = P(z > 2.1) = 1 - .9821 = .0179$$

6.79 The 3000 light bulbs utilized by the manufacturing plant comprise the entire population (that is, this is not a sample from the population) whose length of life is normally distributed with mean $\mu = 500$ and standard deviation $\sigma = 50$. The objective is to find a particular value, x_0, so that

$$P(x \leq x_0) = .01.$$

That is, only 1% of the bulbs will burn out before they are replaced at time x_0. Then

$$P(x \leq x_0) = P(z \leq z_0) = .01 \quad \text{where} \quad z_0 = \frac{x_0 - 500}{50}.$$

From Table 3, the value of z corresponding to an area (in the left tail of the distribution) of .01 is $z_0 = -2.33$. Solving for x_0 corresponding to $z_0 = -2.33$,

$$-2.33 = \frac{x_0 - 500}{50} \quad \Rightarrow \quad -116.5 = x_0 - 500 \quad \Rightarrow \quad x_0 = 383.5$$

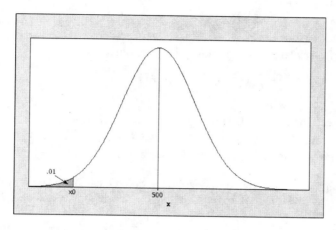

6.81 The random variable of interest is x, the number of persons not showing up for a given flight. This is a binomial random variable with $n = 160$ and $p = P[\text{person does not show up}] = .05$. If there is to be a seat available for every person planning to fly, then there must be at least five persons not showing up. Hence, the probability of interest is $P(x \geq 5)$. Calculate

$$\mu = np = 160(.05) = 8 \quad \text{and} \quad \sigma = \sqrt{npq} = \sqrt{160(.05)(.95)} = \sqrt{7.6} = 2.7$$

Referring to the figure below, a correction for continuity is made to include the entire area under the rectangle associated with the value $x = 5$, and the approximation becomes $P(x \geq 4.5)$. The z-value corresponding to $x = 4.5$ is

$$z = \frac{x - \mu}{\sigma} = \frac{4.5 - 8}{\sqrt{7.6}} = -1.27$$

so that $\qquad P(x \geq 4.5) = P(z \geq -1.27) = 1 - .1020 = .8980$

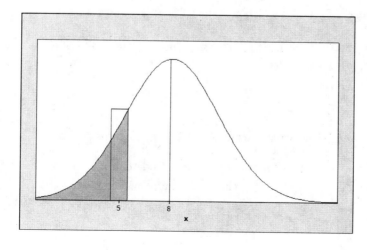

6.83 **a** Let x be the number of plants with red petals. Then x has a binomial distribution with $n = 100$ and $p = .75$.

b Since $np = 100(.75) = 75$ and $nq = 100(.25)$ are both greater than 5, the normal approximation is appropriate.

c Calculate

$$\mu = np = 100(.75) = 75 \quad \text{and} \quad \sigma = \sqrt{npq} = \sqrt{100(.75)(.25)} = \sqrt{18.75} = 4.33$$

A correction for continuity is made to include the entire area under the rectangles corresponding to $x = 70$ and $x = 80$. Hence the approximation will be

$$P(69.5 < x < 80.5) = P\left(\frac{69.5 - 75}{4.33} < z < \frac{80.5 - 75}{4.33}\right) = P(-1.27 < z < 1.27) = .8980 - .1020 = .7960$$

d The probability that 53 or fewer plants have red flower is approximated as

$$P(x < 53.5) = P\left(z < \frac{53.5 - 75}{4.33}\right) = P(z < -4.97) \approx 0$$

This would be considered and unusual event.

e If the value $p = .75$ is correct, the only explanation for the unusual occurrence in part **d** is that the $n = 100$ seeds do not represent a random sample from the population of peony plants. Perhaps the sample became contaminated in some way; some other uncontrolled variable is affecting the flower color.

6.85 For the binomial random variable x, the mean and standard deviation are calculated under the assumption that there is no difference between the effect of TV and reading on calorie intake, and hence that $p = .5$. Then

$$\mu = np = 30(.5) = 15 \quad \text{and} \quad \sigma = \sqrt{npq} = \sqrt{30(.5)(.5)} = 2.7386$$

If there is no difference between TV and reading, the z-score for the observed value of x, $x = 19$, is

$$z = \frac{x - \mu}{\sigma} = \frac{19 - 15}{2.7386} = 1.461$$

That is, the observed value lies 1.461 standard deviations above the mean. This is not an unlikely occurrence. Hence, we would have no reason to believe that there is a difference between calorie intake for TV watchers versus readers.

6.87 The random variable y, the percentage of tax returns audited, has a normal distribution with $\mu = 1.55$ and $\sigma = .45$.

a $P(y > 2) = P\left(z > \frac{2 - 1.55}{.45}\right) = P(z > 1) = 1 - .8413 = .1587$

b Define x to be the number of states in which more than 2% of its returns were audited. Then x has a binomial distribution with

$$\mu = np = 50(.1587) = 7.935 \quad \text{and} \quad \sigma = \sqrt{npq} = \sqrt{50(.1587)(.8413)} = 2.583$$

The expected value of x is $E(x) = \mu = 7.935$.

c The value $x = 15$ has z-score

$$z = \frac{x - \mu}{\sigma} = \frac{15 - 7.935}{2.583} = 2.73$$

It is unlikely that as many as 15 states will have more than 2% of its returns audited.

6.89 The scores are approximately normal with mean $\mu = 75$ and standard deviation $\sigma = 12$. We need to find a value of x, say $x = c$, such that $P(x > c) = .15$. The z-value corresponding to $x = c$ is

$$z = \frac{x - \mu}{\sigma} = \frac{c - 75}{12}$$

From Table 3, the z-value corresponding to an area of .15 in the right tail of the normal distribution is a value that has area .8500 to its left. The closest value given in the table is $z = 1.04$. Then,

$$\frac{c - 75}{12} = 1.04 \Rightarrow c = 75 + 1.04(12) = 87.48$$

The proper score to designate "extroverts" would be any score higher than 87.48.

6.91 The measurements are approximately normal with mean $\mu = 39.83$ and standard deviation $\sigma = 2.05$.

a

$$P(36.5 < x < 43.5) = P\left(\frac{36.5 - 39.83}{2.05} < z < \frac{43.5 - 39.83}{2.05}\right) = P(-1.62 < z < 1.79) = .9633 - .0526 = .9107$$

b For the standard normal z, we know that $P(-1.96 < z < 1.96) = .9750 - .0250 = .9500$. Substituting for z in the probability statement, we write

$$P(-1.96 < \frac{x - \mu}{\sigma} < 1.96) \Rightarrow P(\mu - 1.96\sigma < x < \mu + 1.96\sigma) = .95$$

That is, 95% of the chest measurements, x, will lie between $\mu - 1.96\sigma$ and $\mu + 1.96\sigma$, or

$$39.83 \pm 1.96(2.05) \text{ which is } 35.812 \text{ to } 43.848.$$

c Refer to the data distribution given in Exercise 6.90, and count the number of measurements falling in each of the two intervals.

- Between 36.5 and 43.5: $\dfrac{420 + 749 + \ldots + 370}{5738} = \dfrac{5283}{5738} = .921$ (compared to .9107)

- Between 35.812 and 43.848: $\dfrac{185 + 420 + \ldots + 370}{5738} = \dfrac{5468}{5738} = .953$ (compared to .95)

6.93 **a-b** The random variable x has a binomial distribution with $n = 100$ and $p = .51$. Calculate $\mu = np = 100(.51) = 51$ and $\sigma = \sqrt{npq} = \sqrt{100(.51).49} = 4.999$. The probability that x is 60 or more is approximated as

$$P(x \geq 59.5) = P(z \geq \frac{59.5 - 51}{4.999}) = P(z \geq 1.70) = 1 - .9554 = .0446$$

6.95 **a** It is given that the scores on a national achievement test were approximately normally distributed with a mean of 540 and standard deviation of 110. It is necessary to determine how far, in standard deviations, a score of 680 departs from the mean of 540. Calculate

$$z = \frac{x - \mu}{\sigma} = \frac{680 - 540}{110} = 1.27 .$$

b To find the percentage of people who scored higher than 680, we find the area under the standardized normal curve greater than 1.27. Using Table 3, this area is equal to

$$P(x > 680) = P(z > 1.27) = 1 - .8980 = .1020$$

Thus, approximately 10.2% of the people who took the test scored higher than 680.

6.97 It is given that the probability of a successful single transplant from the early gastrula stage is .65. In a sample of 100 transplants, the mean and standard deviation of the binomial distribution are

$$\mu = np = 100(.65) = 65 \quad \text{and} \quad \sigma = \sqrt{npq} = \sqrt{100(.65)(.35)} = 4.770$$

It is necessary to find the probability that more than 70 transplants will be successful. This is approximated by the area under a normal curve with $\mu = 65$ and $\sigma = 4.77$ to the right of 70.5. The z-value corresponding

to $x = 70.5$ is $\quad z = \dfrac{x - \mu}{\sigma} = \dfrac{70.5 - 65}{4.77} = 1.15$ and

$$P(x > 70) \approx P(z > 1.15) = 1 - .8749 = .1251$$

81

7: Sampling Distributions

7.1 You can select a simple random sample of size $n = 20$ using Table 10 in Appendix I. First choose a starting point and consider the first three digits in each number. Since the experimental units have already been numbered from 000 to 999, the first 20 can be used. The three digits OR the (three digits – 500) will identify the proper experimental unit. For example, if the three digits are 742, you should select the experimental unit numbered $742 - 500 = 242$. The probability that any three digit number is selected is $2/1000 = 1/500$. One possible selection for the sample size $n = 20$ is

242	134	173	128	399
056	412	188	255	388
469	244	332	439	101
399	156	028	238	231

7.3 Each student will obtain a different sample, using Table 10 in Appendix I.

7.5 If all of the town citizenry is likely to pass this corner, a sample obtained by selecting every tenth person is probably a fairly random sample.

7.7 Voter registration lists and DMV records will systematically exclude a large segment of the general population – nondrivers, people who do not own cars, nonvoters and so on.

7.9 Use a randomization scheme similar to that used in Exercise 7.1. Number each of the 100 individuals from 00 to 99. To choose the 50 individuals who will receive the experimental treatment, select 50 two-digit random numbers from Table 10. Each two-digit number will identify the proper experimental unit. The other 50 individuals will be placed in the control group.

7.11 **a** The sample was chosen from Native American youth attending an after-school program in Minneapolis, MN who were willing to participate. This sample is not randomly selected; it is a convenience sample.
b Valid inferences can be made from this study *only if* the convenience sample chosen by the researcher *behaves like a random sample*. That is, the in this particular after-school program must be representative of the population of Native American youth as a whole.
c In order to increase the chances of obtaining a sample that is representative of the population of Native American youth as a whole, the researcher might try to obtain a larger base of students to choose from. Perhaps there is a computerized database from which he or she might select a random sample.

7.13 **a** The first question is more unbiased.
b Notice that the percentage favoring the new space program drops dramatically when the phrase "spending billions of dollars" is added to the question.

7.14 Answers will vary. Many of the questions used in this policy survey were biased towards the political views held by members of the Republican Party.

7.15 Regardless of the shape of the population from which we are sampling, the sampling distribution of the sample mean will have a mean μ equal to the mean of the population from which we are sampling, and a standard deviation equal to $\sigma/\sqrt{n}$.

a $\mu = 10$; $\sigma/\sqrt{n} = 3/\sqrt{36} = .5$
b $\mu = 5$; $\sigma/\sqrt{n} = 2/\sqrt{100} = .2$
c $\mu = 120$; $\sigma/\sqrt{n} = 1/\sqrt{8} = .3536$

7.17 **a** The probability histogram is shown on the next page. It is far from mound-shaped.

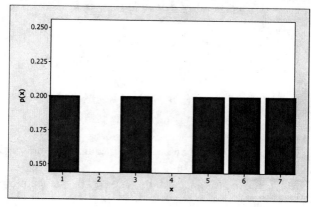

b Answers will vary from student to student.
c Using the data given in the text as an example, the relative frequency histogram shows a clearly mound-shape.

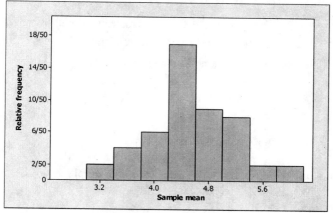

7.19 For a population with $\sigma = 1$, the standard error of the mean is
$$\sigma/\sqrt{n} = 1/\sqrt{n}$$
The values of $\sigma/\sqrt{n}$ for various values of n are tabulated below. Notice that the standard error *decreases* as the sample size *increases*.

n	1	2	4	9	16	25	100
$SE(\overline{x}) = \sigma/\sqrt{n}$	1.00	.707	.500	.333	.250	.200	.100

7.21 **a** Since $n = 49$, the Central Limit Theorem is used. The sampling distribution of $\overline{x}$ will be approximately normal.
b The mean of the sampling distribution of $\overline{x}$ is $\mu = 53$ and the standard deviation (or standard error) is $\sigma/\sqrt{n} = 21/\sqrt{9} = 3$.

7.23 **a** Since $n = 40$, the Central Limit Theorem is used. The sampling distribution of $\overline{x}$ will be approximately normal.
b The mean of the sampling distribution of $\overline{x}$ is $\mu = 100$ and the standard deviation (or standard error) is $\sigma/\sqrt{n} = 20/\sqrt{40} = 3.16$.

7.25   **a**   If the sample population is normal, the sampling distribution of $\bar{x}$ will also be normal (regardless of the sample size) with mean $\mu = 106$ and standard deviation (or *standard error*) given as

$$\sigma/\sqrt{n} = 12/\sqrt{25} = 2.4$$

b   Calculate $z = \dfrac{\bar{x} - \mu}{\sigma/\sqrt{n}} = \dfrac{110 - 106}{2.4} = 1.67$, so that

$$P(\bar{x} > 110) = P(z > 1.67) = 1 - .9525 = .0475$$

c $P(102 < \bar{x} < 110) = P(-1.67 < z < 1.67) = .9525 - .0475 = .9050$

7.27   **a**   Age of equipment, technician error, technician fatigue, equipment failure, difference in chemical purity, contamination from outside sources, and so on.

b   The variability in the average measurement is measured by the standard error, $\sigma/\sqrt{n}$. In order to decrease this variability you should increase the sample size n.

7.29   The number of bacteria in one cubic foot of water can be thought of as the sum of 1728 random variables, each of which is the number of bacteria in a particular cubic inch of water. Hence, the Central Limit Theorem insures the approximate normality of the sum.

7.31   **a**   The random variable $T = \sum x_i$, were x_i is normally distributed with mean $\mu = 422$ and standard deviation $\sigma = 13$ for $i = 1, 2, 3$. The Central Limit Theorem states that T is normally distributed with mean $n\mu = 3(422) = 1266$ and standard deviation $\sigma\sqrt{n} = 13\sqrt{3} = 22.5167$.

b   Calculate

$$z = \frac{T - 1266}{22.5167} = \frac{1300 - 1266}{22.5167} = 1.51.$$

Then $P(T > 1300) = P(z < 1.51) = 1 - .9345 = .0655$.

7.33   **a**   Since the original population is normally distributed, the sample mean $\bar{x}$ is also normally distributed (for any sample size) with mean μ and standard deviation

$$\sigma/\sqrt{n} = 0.8/\sqrt{130} = .07016$$

The z-value corresponding to $\bar{x} = 98.25$ is

$$z = \frac{\bar{x} - \mu}{\sigma/\sqrt{n}} = \frac{98.25 - 98.6}{0.8/\sqrt{130}} = -4.99$$

and

$$P(\bar{x} < 98.25) = P(z < -4.99) \approx 0$$

b   Since the probability is extremely small, the average temperature of 98.25 degrees is very unlikely.

7.35   **a** $p = .3; \ SE(\hat{p}) = \sqrt{\dfrac{pq}{n}} = \sqrt{\dfrac{.3(.7)}{100}} = .0458$

b $p = .1; \ SE(\hat{p}) = \sqrt{\dfrac{pq}{n}} = \sqrt{\dfrac{.1(.9)}{400}} = .015$

c $p = .6; \ SE(\hat{p}) = \sqrt{\dfrac{pq}{n}} = \sqrt{\dfrac{.6(.4)}{250}} = .0310$

7.37   **a**   Since $\hat{p}$ is approximately normal, with standard deviation $SE(\hat{p}) = \sqrt{\dfrac{pq}{n}} = \sqrt{\dfrac{.4(.6)}{75}} = .0566$, the probability of interest is

$$P(\hat{p} \le .43) = P\left(z \le \frac{.43 - .4}{.0566}\right) = P(z \le .53) = .7019$$

84

b The probability is approximated as

$$P(.35 \le \hat{p} \le .43) = P\left[\frac{.35-.4}{.0566} \le z \le \frac{.43-.4}{.0566}\right]$$
$$= P(-.88 \le z \le .53) = .7019 - .1894 = .5125$$

7.39 The values $SE = \sqrt{pq/n}$ for $n = 100$ and various values of p are tabulated and graphed below. Notice that SE is a maximum for $p = .5$ and becomes very small for p near zero and one.

p	.01	.10	.30	.50	.70	.90	.99
$SE(\hat{p})$	.0099	.03	.0458	.05	.0458	.03	.0099

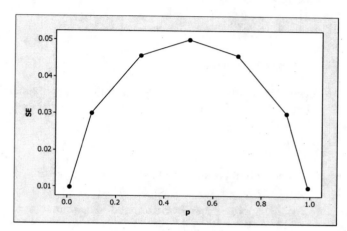

7.41 **a** Since $np = 20$ and $nq = 60$, the normal approximation is appropriate. The sampling distribution of $\hat{p}$ will be approximately normal.

b The mean of the sampling distribution of $\hat{p}$ is $p = .25$ and the standard deviation (or standard error) is

$$\sqrt{\frac{pq}{n}} = \sqrt{\frac{.25(.75)}{80}} = .04841.$$

c The probability of interest is $P(.18 < \hat{p} < .44)$. When $\hat{p} = .18$ and $\hat{p} = .44$,

$$z = \frac{\hat{p}-p}{\sqrt{\frac{pq}{n}}} = \frac{.18-.25}{.04841} = -1.45 \text{ and } z = \frac{\hat{p}-p}{\sqrt{\frac{pq}{n}}} = \frac{.44-.25}{.04841} = 3.92$$

Then $P(.18 < \hat{p} < .44) = P(-1.45 < z < 3.92) = 1 - .0735 = .9265$

7.43 **a** For $n = 100$ and $p = .78$, $np = 78$ and $nq = 22$ are both greater than 5. Therefore, the normal approximation will be appropriate, with mean $p = .78$ and $SE = \sqrt{\frac{pq}{n}} = \sqrt{\frac{.78(.22)}{100}} = .0414$.

b $P(\hat{p} < .75) = P\left(z < \frac{.75-.78}{.0414}\right) = P(z < -.72) = .2358$

c $P(.7 < \hat{p} < .75) = P\left(\frac{.7-.78}{.0414} < z < \frac{.75-.78}{.0414}\right) = P(-1.93 < z < -.72) = .2358 - .0268 = .2090$

d The value $\hat{p} = .65$ lies

85

$$z = \frac{\hat{p} - p}{\sqrt{\dfrac{pq}{n}}} = \frac{.65 - .78}{.0414} = -3.14$$

standard deviations from the mean. Also, $P(\hat{p} \le .65) = P(z \le -3.14) = .0008$. This is an unlikely occurrence, assuming that $p = .78$. Perhaps the sampling was not random, or the 78% figure is not correct.

7.45 **a** The random variable $\hat{p}$, the sample proportion of brown M&Ms in a package of $n = 55$, has a binomial distribution with $n = 55$ and $p = .13$. Since $np = 7.15$ and $nq = 47.85$ are both greater than 5, this binomial distribution can be approximated by a normal distribution with mean $p = .13$ and

$$SE = \sqrt{\frac{.13(.87)}{55}} = .04535.$$

b $P(\hat{p} < .2) = P\left(z < \dfrac{.2 - .13}{.04535}\right) = P(z < 1.54) = .9382$

c $P(\hat{p} > .35) = P\left(z > \dfrac{.35 - .13}{.04535}\right) = P(z > 4.85) \approx 1 - 1 = 0$

d From the Empirical Rule (and the general properties of the normal distribution), approximately 95% of the measurements will lie within 2 (or 1.96) standard deviations of the mean:
$$p \pm 2SE \implies .13 \pm 2(.04535)$$
$$.13 \pm .09 \quad \text{or} \quad .04 \text{ to } .22$$

7.47 **a** The random variable $\hat{p}$, the sample proportion of consumers who like nuts or caramel in their chocolate, has a binomial distribution with $n = 200$ and $p = .75$. Since $np = 150$ and $nq = 50$ are both greater than 5, this binomial distribution can be approximated by a normal distribution with mean $p = .75$ and $SE = \sqrt{\dfrac{.75(.25)}{200}} = .03062$.

b $P(\hat{p} > .80) = P\left(z > \dfrac{.80 - .75}{.03062}\right) = P(z > 1.63) = 1 - .9484 = .0516$

c From the Empirical Rule (and the general properties of the normal distribution), approximately 95% of the measurements will lie within 2 (or 1.96) standard deviations of the mean:
$$p \pm 2SE \implies .75 \pm 2(.03062)$$
$$.75 \pm .06 \quad \text{or} \quad .69 \text{ to } .81$$

7.49 **a** The upper and lower control limits are
$$UCL = \bar{\bar{x}} + 3\frac{s}{\sqrt{n}} = 155.9 + 3\frac{4.3}{\sqrt{5}} = 155.9 + 5.77 = 161.67$$
$$LCL = \bar{\bar{x}} - 3\frac{s}{\sqrt{n}} = 155.9 - 3\frac{4.3}{\sqrt{5}} = 155.9 - 5.77 = 150.13$$

b The control chart is constructed by plotting two horizontal lines, one the upper control limit and one the lower control limit (see Figure 7.13 in the text). Values of $\bar{x}$ are plotted, and should remain within the control limits. If not, the process should be checked.

7.51 **a** The upper and lower control limits for a p chart are
$$UCL = \bar{p} + 3\sqrt{\frac{\bar{p}(1 - \bar{p})}{n}} = .035 + 3\sqrt{\frac{.035(.965)}{100}} = .035 + .055 = .090$$
$$LCL = \bar{p} - 3\sqrt{\frac{\bar{p}(1 - \bar{p})}{n}} = .035 - 3\sqrt{\frac{.035(.965)}{100}} = .035 - .055 = -.020$$

or $LCL = 0$ (since p cannot be negative).

b The control chart is constructed by plotting two horizontal lines, one the upper control limit and one the lower control limit (see Figure 7.14 in the text). Values of $\hat{p}$ are plotted, and should remain within the control limits. If not, the process should be checked.

7.53   **a**   The upper and lower control limits are

$$UCL = \bar{\bar{x}} + 3\frac{s}{\sqrt{n}} = 10,752 + 3\frac{1605}{\sqrt{5}} = 10,752 + 2153.3 = 12,905.3$$

$$LCL = \bar{\bar{x}} - 3\frac{s}{\sqrt{n}} = 10,752 - 3\frac{1605}{\sqrt{5}} = 10,752 - 2153.3 = 8598.7$$

b   The $\bar{x}$ chart will allow the manager to monitor daily gains or losses to see whether there is a problem with any particular table.

7.55   Calculate $\bar{p} = \dfrac{\sum \hat{p}_i}{k} = \dfrac{.14 + .21 + \cdots + .26}{30} = .197$. The upper and lower control limits for the p chart are then

$$UCL = \bar{p} + 3\sqrt{\frac{\bar{p}(1 - \bar{p})}{n}} = .197 + 3\sqrt{\frac{.197(.803)}{100}} = .197 + .119 = .316$$

$$LCL = \bar{p} - 3\sqrt{\frac{\bar{p}(1 - \bar{p})}{n}} = .197 - 3\sqrt{\frac{.197(.803)}{100}} = .197 - .119 = .078$$

7.57   Using all 104 measurements, the value of s is calculated to be $s = .006717688$ and $\bar{\bar{x}} = .0256$. Then the upper and lower control limits are

$$UCL = \bar{\bar{x}} + 3\frac{s}{\sqrt{n}} = .0256 + 3\frac{.06717688}{\sqrt{4}} = .0357$$

$$LCL = \bar{\bar{x}} - 3\frac{s}{\sqrt{n}} = .0256 - 3\frac{.06717688}{\sqrt{4}} = .0155$$

7.59   Refer to Exercise 7.58, where the upper and lower control limits were calculated as $LCL = 31.423$ and $UCL = 31.977$. The sample mean is outside the control limits in hours 2, 3 and 4. The process should be checked.

7.61   Refer to Exercise 7.60. If samples of size $n = 3$ are drawn without replacement, there are 4 possible samples with sample means shown below.

Sample	Observations	$\bar{x}$
1	6, 1, 3	3.333
2	6, 1, 2	3
3	6, 3, 2	3.667
4	1, 3, 2	2

The sampling distribution of $\bar{x}$ is then

$$p(\bar{x}) = \frac{1}{4} \quad \text{for } \bar{x} = 2, 3, 3.333, 3.667$$

The sampling distribution is shown below.

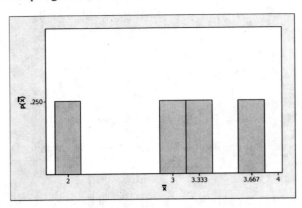

7.63 Since $n = 400$, the Central Limit Theorem ensures that the sampling distribution of $\bar{x}$ will be approximately normal with mean $\mu = 1110$ and standard deviation $\sigma / \sqrt{n} = 80 / \sqrt{400} = 4$.

a $P(1100 < \bar{x} < 1110) = P(\dfrac{1100 - 1110}{4} < z < \dfrac{1110 - 1110}{4}) = P(-2.5 < z < 0) = .5000 - .0062 = .4938.$

b $P(\bar{x} > 1120) = P(z > \dfrac{1120 - 1110}{4}) = P(z > 2.5) = 1 - .9938 = .0062.$

c $P(\bar{x} < 900) = P(z < \dfrac{900 - 1110}{4}) = P(z < -52.5) = .0000.$

7.65 **a** Using the range approximation, the standard deviation σ can be approximated as
$$\sigma \approx \frac{R}{4} = \frac{55 - 5}{4} = 12.5$$

b The sampling distribution of $\bar{x}$ is approximately normal with mean μ and standard error
$$\sigma / \sqrt{n} \approx 12.5 / \sqrt{400} = .625$$
Then $P(|\bar{x} - \mu| \le 2) = P\left(\dfrac{-2}{.625} \le z \le \dfrac{2}{.625} \right) = P(-3.2 \le z \le 3.2) = .9993 - .0007 = .9986$.

c If the scientists are worried that the estimate of $\mu = 35$ is too high, and if your estimate is $\bar{x} = 31.75$, then your estimate lies
$$z = \frac{\bar{x} - \mu}{\sigma / \sqrt{n}} = \frac{31.75 - 35}{.625} = -5.2$$

standard deviations below the mean. This is a very unlikely event if in fact $\mu = 35$. It is more likely that the scientists are correct in assuming that the mean is an overestimate of the mean biomass for tropical woodlands.

7.67 **a** To divide a group of 20 people into two groups of 10, use Table 10 in Appendix I. Assign an identification number from 01 to 20 to each person. Then select ten two digit numbers from the random number table to identify the ten people in the first group. (If the number is greater than 20, subtract multiples of 20 from the random number until you obtain a number between 01 and 20.)
b Although it is not possible to select an actual random sample from this hypothetical population, the researcher must obtain a sample that *behaves like* a random sample. A large database of some sort should be used to ensure a fairly representative sample.
c The researcher has actually selected a *convenience sample*; however, it will probably behave like a simple random sample, since a person's enthusiasm for a paid job should not affect his response to this psychological experiment.

7.69 **a** From the 50 lettuce seeds, the researcher must choose a group of 26 and a group of 13 for the experiment. Identify each seed with a number from 01 to 50 and then select random numbers from Table 10. The first 26 numbers chosen will identify the seeds in the first Petri dish, and the next 13 will identify the seeds in the third Petri dish. If the same number is picked twice, simply ignore it and go on to the next number. Use a similar procedure to choose the groups of 26 and 13 radish seeds.
b The seeds in these two packages must be representative of all seeds in the general population of lettuce and radish seeds.

7.71 Referring to Table 10 in Appendix I, we will select 20 numbers. First choose a starting point and consider the first four digits in each number. If the four digits are a number greater than 7000, discard it. Continue until 20 numbers have been chosen. The customers have already been numbered from 0001 to 7000. One possible selection for the sample size $n = 20$ is

1048	2891	5108	4866
2236	6355	0236	5416
2413	0942	0101	3263
4216	1036	5216	2933
3757	0711	0705	0248

7.73 **a** Since each cluster (a city block) is censused, this is an example of cluster sampling.
b This is a 1-in-10 systematic sample.
c The wards are the strata, and the sample is a stratified sample.
d This is a 1-in-10 systematic sample.
e This is a simple random sample from the population of all tax returns filed in the city of San Bernardino, California.

7.75 **a** The number of packages which can be assembled in 8 hours is the sum of 8 observations on the random variable described here. Hence, its mean is $n\mu = 8(16.4) = 131.2$ and its standard deviation is $\sigma\sqrt{n} = 1.3\sqrt{8} = 3.677$.
b If the original population is approximately normal, the sampling distribution of a sum of 8 normal random variables will also be approximately normal. Since the original population is exactly normal, so will be the sampling distribution of the sum.
c $P(x > 135) = P\left(z > \dfrac{135 - 131.2}{3.677}\right) = P(z > 1.03) = 1 - .8485 = .1515$

7.77 **a** The average proportion of defectives is

$$\overline{p} = \frac{.04 + .02 + \cdots + .03}{25} = .032$$

and the control limits are

$$UCL = \overline{p} + 3\sqrt{\frac{\overline{p}(1 - \overline{p})}{n}} = .032 + 3\sqrt{\frac{.032(.968)}{100}} = .0848$$

and

$$LCL = \overline{p} - 3\sqrt{\frac{\overline{p}(1 - \overline{p})}{n}} = .032 - 3\sqrt{\frac{.032(.968)}{100}} = -.0208$$

If subsequent samples do not stay within the limits, $UCL = .0848$ and $LCL = 0$, the process should be checked.
b From part **a**, we must have $\hat{p} > .0848$.
c An erroneous conclusion will have occurred if in fact $p < .0848$ and the sample has produced $\hat{p} = .15$ by chance. One can obtain an upper bound on the probability of this particular type of error by calculating $P(\hat{p} \geq .15$ when $p = .0848)$.

7.79 Refer to Exercise 7.77, in which $UCL = .0848$ and $LCL = 0$. For the next 5 samples, the values of $\hat{p}$ are .02, .04, .09, .07, .11. Hence, samples 3 and 5 are producing excess defectives. The process should be checked.

89

7.81   Answers will vary from student to student. Paying cash for opinions will not necessarily produce a random sample of opinions of all Pepsi and Coke drinkers.

7.83   **a**   Since the fill per can is normal with mean 12 and standard deviation 0.2, the total fill for a pack of six cans will also have a normal distribution with mean $n\mu = 24(12) = 288$ and standard deviation

$$\sigma\sqrt{n} = 0.2\sqrt{24} = .9798$$

   b   Let T be the total fill for the case of soda. Then

$$P(T < 286) = P(z < \frac{286 - 288}{.9798}) = P(z < -2.04) = .0207$$

   c $P(\bar{x} < 11.8) = P(z < \frac{11.8 - 12}{0.2/\sqrt{6}}) = P(z < -2.45) = .0071$

7.85   **a**   The average proportion of inoperable components is

$$\bar{p} = \frac{6 + 7 + \cdots + 5}{50(15)} = \frac{75}{750} = .10$$

and the control limits are

$$UCL = \bar{p} + 3\sqrt{\frac{\bar{p}(1-\bar{p})}{n}} = .10 + 3\sqrt{\frac{.10(.90)}{50}} = .2273$$

and $$LCL = \bar{p} - 3\sqrt{\frac{\bar{p}(1-\bar{p})}{n}} = .10 - 3\sqrt{\frac{.1(.9)}{50}} = -.0272$$

If subsequent samples do not stay within the limits, $UCL = .2273$ and $LCL = 0$, the process should be checked.

8: Large-Sample Estimation

8.1 The margin of error in estimation provides a practical upper bound to the difference between a particular estimate and the parameter which it estimates. In this chapter, the margin of error is $1.96 \times$ (standard error of the estimator).

8.3 For the estimate of μ given as $\bar{x}$, the margin of error is $1.96 \, SE = 1.96 \dfrac{\sigma}{\sqrt{n}}$.

    a $1.96\sqrt{\dfrac{0.2}{30}} = .160$ **b** $1.96\sqrt{\dfrac{0.9}{30}} = .339$ **c** $1.96\sqrt{\dfrac{1.5}{30}} = .438$

8.5 The margin of error is $1.96 \, SE = 1.96 \dfrac{\sigma}{\sqrt{n}}$, where σ can be estimated by the sample standard deviation s for large values of n.

    a $1.96\sqrt{\dfrac{4}{50}} = .554$ **b** $1.96\sqrt{\dfrac{4}{500}} = .175$ **c** $1.96\sqrt{\dfrac{4}{5000}} = .055$

8.7 For the estimate of p given as $\hat{p} = x/n$, the margin of error is $1.96 \, SE = 1.96\sqrt{\dfrac{pq}{n}}$.

    a $1.96\sqrt{\dfrac{(.5)(.5)}{30}} = .179$ **b** $1.96\sqrt{\dfrac{(.5)(.5)}{100}} = .098$

    c $1.96\sqrt{\dfrac{(.5)(.5)}{400}} = .049$ **d** $1.96\sqrt{\dfrac{(.5)(.5)}{1000}} = .031$

8.9 For the estimate of p given as $\hat{p} = x/n$, the margin of error is $1.96 \, SE = 1.96\sqrt{\dfrac{pq}{n}}$. Use the value given in the exercise for p.

    a $1.96\sqrt{\dfrac{(.1)(.9)}{100}} = .0588$ **b** $1.96\sqrt{\dfrac{(.3)(.7)}{100}} = .0898$ **c** $1.96\sqrt{\dfrac{(.5)(.5)}{100}} = .098$

    d $1.96\sqrt{\dfrac{(.7)(.3)}{100}} = .0898$ **e** $1.96\sqrt{\dfrac{(.9)(.1)}{100}} = .0588$

    f The largest margin of error occurs when $p = .5$.

8.11 The point estimate for p is given as $\hat{p} = \dfrac{x}{n} = \dfrac{655}{900} = .728$ and the margin of error is approximately

$$1.96\sqrt{\dfrac{\hat{p}\hat{q}}{n}} = 1.96\sqrt{\dfrac{.728(.272)}{900}} = .029$$

8.13 The point estimate for p is given as $\hat{p} = \dfrac{x}{n} = \dfrac{450}{500} = .90$ and the margin of error is approximately

$$1.96\sqrt{\dfrac{\hat{p}\hat{q}}{n}} = 1.96\sqrt{\dfrac{.90(.10)}{500}} = .0263$$

8.15 The point estimate of μ is $\bar{x} = 39.8°$ and the margin of error with $s = 17.2$ and $n = 50$ is

$$1.96 \, SE = 1.96\dfrac{\sigma}{\sqrt{n}} \approx 1.96\dfrac{s}{\sqrt{n}} = 1.96\dfrac{17.2}{\sqrt{50}} = 4.768$$

91

8.17   The point estimate of μ is $\bar{x} = 7.2\%$ and the margin of error with $s = 5.6\%$ and $n = 200$ is

$$1.96\ SE = 1.96\frac{\sigma}{\sqrt{n}} \approx 1.96\frac{s}{\sqrt{n}} = 1.96\frac{5.6}{\sqrt{200}} = .776$$

8.19   **a**   The point estimate for p is given as $\hat{p} = \frac{x}{n} = .75$ and the margin of error is approximately

$$1.96\sqrt{\frac{\hat{p}\hat{q}}{n}} = 1.96\sqrt{\frac{.75(.25)}{1004}} = .0268$$

b   The sampling error should be reported by using the maximum margin of error using $p = .5$, and by rounding off to the nearest percent:

$$1.96\sqrt{\frac{\hat{p}\hat{q}}{n}} = 1.96\sqrt{\frac{.5(.5)}{1004}} = .0309 \quad \text{or} \quad \pm 3.1\%$$

The margin of error given in the article is not correct, unless the authors chose to round up to the next "half-percentage" point.

8.21   **a**   This method of sampling would not be random, since only interested viewers (those who were adamant in their approval or disapproval) would reply.
b   The results of such a survey will not be valid, and a margin or error would be useless, since its accuracy is based on the assumption that the sample was random.

8.23   A point estimate for the mean length of time is $\bar{x} = 19.3$, with margin of error

$$1.96\ SE = 1.96\frac{\sigma}{\sqrt{n}} \approx 1.96\frac{s}{\sqrt{n}} = 1.96\frac{5.2}{\sqrt{30}} = 1.86$$

8.25   A 90% confidence interval for the population mean μ is given by

$$\bar{x} \pm 1.645\frac{\sigma}{\sqrt{n}}$$

where σ can be estimated by the sample standard deviation s for large values of n.

a $.84 \pm 1.645\sqrt{\dfrac{.086}{125}} = .84 \pm .043 \quad$ or $\quad .797 < \mu < .883$

b $21.9 \pm 1.645\sqrt{\dfrac{3.44}{50}} = 21.9 \pm .431 \quad$ or $\quad 21.469 < \mu < 22.331$

c   Intervals constructed in this manner will enclose the true value of μ 90% of the time in repeated sampling. Hence, we are fairly confident that these particular intervals will enclose μ.

8.27   Calculate $\hat{p} = \dfrac{x}{n} = \dfrac{263}{300} = .877$. Then an approximate 90% confidence interval for p is

$$\hat{p} \pm 1.645\sqrt{\frac{\hat{p}\hat{q}}{n}} = .877 \pm 1.645\sqrt{\frac{.877(.123)}{300}} = .877 \pm .031$$

or $.846 < p < .908$.

8.29   The width of a 95% confidence interval for μ is given as $1.96\dfrac{\sigma}{\sqrt{n}}$. Hence,

a   When $n = 100$, the width is $2\left(1.96\dfrac{10}{\sqrt{100}}\right) = 2(1.96) = 3.92$.

b   When $n = 200$, the width is $2\left(1.96\dfrac{10}{\sqrt{200}}\right) = 2(1.386) = 2.772$.

92

c When $n = 400$, the width is $2\left(1.96\dfrac{10}{\sqrt{400}}\right) = 2(.98) = 1.96$.

8.31 **a** A 90% confidence interval for μ is $\bar{x} \pm 1.645\dfrac{\sigma}{\sqrt{n}}$. Hence, its width is

$$2\left(1.645\dfrac{\sigma}{\sqrt{n}}\right) = 2\left(1.645\dfrac{10}{\sqrt{100}}\right) = 2(1.645) = 3.29$$

b A 99% confidence interval for μ is $\bar{x} \pm 2.58\dfrac{\sigma}{\sqrt{n}}$. Hence, its width is

$$2\left(2.58\dfrac{\sigma}{\sqrt{n}}\right) = 2\left(2.58\dfrac{10}{\sqrt{100}}\right) = 2(2.58) = 5.16$$

c Notice that as the confidence coefficient increases, so does the width of the confidence interval. If we wish to be more confident of enclosing the unknown parameter, we must make the interval wider.

8.33 With $n = 40$, $\bar{x} = 3.7$ and $s = .5$ and $\alpha = .01$, a 99% confidence interval for μ is approximated by

$$\bar{x} \pm 2.58\dfrac{s}{\sqrt{n}} = 3.7 \pm 2.58\dfrac{.5}{\sqrt{40}} = 3.7 \pm .204 \text{ or } 3.496 < \mu < 3.904$$

In repeated sampling, 99% of all intervals constructed in this manner will enclose μ. Hence, we are fairly certain that this particular interval contains μ. (In order for this to be true, the sample must be randomly selected.)

8.35 **a** With $n = 35$, $\bar{x} = 1.01$ and $s = .18$ and $\alpha = .01$, a 99% confidence interval for μ is approximated by

$$\bar{x} \pm 2.58\dfrac{s}{\sqrt{n}} = 1.01 \pm 2.58\dfrac{.18}{\sqrt{35}} = 1.01 \pm .078 \text{ or } .932 < \mu < 1.088$$

b In repeated sampling, 99% of all intervals constructed in this manner will enclose μ. Hence, we are fairly certain that this particular interval contains μ. (In order for this to be true, the sample must be randomly selected.)

c No. Since the value $\mu = 1$ is contained in the interval in part **a**, it is one of several possible values for μ. The quality control department would have no reason to be concerned that the trays are being over- or underfilled.

8.37 **a** The point estimate of p is $\hat{p} = \dfrac{x}{n} = \dfrac{68}{500} = .136$, and the approximate 95% confidence interval for p is

$$\hat{p} \pm 1.96\sqrt{\dfrac{\hat{p}\hat{q}}{n}} = .136 \pm 1.96\sqrt{\dfrac{.136(.864)}{500}} = .136 \pm .030$$

or $.106 < p < .166$.

b In order to increase the accuracy of the confidence interval, you must decrease its width. You can accomplish this by (1) increasing the sample size n, or (2) decreasing $z_{\alpha/2}$ by decreasing the confidence coefficient.

8.39 **a** The 99% confidence interval for μ is

$$\bar{x} \pm 2.58\dfrac{s}{\sqrt{n}} = 98.25 \pm 2.58\dfrac{0.73}{\sqrt{130}} = 98.25 \pm .165 \text{ or } 98.085 < \mu < 98.415$$

b Since the possible values for μ given in the confidence interval does not include the value $\mu = 98.6$, it is not likely that the true average body temperature for healthy humans is 98.6, the usual average temperature cited by physicians and others.

8.41 **a** When estimating the difference $\mu_1 - \mu_2$, the $(1-\alpha)100\%$ confidence interval is

$(\overline{x}_1 - \overline{x}_2) \pm z_{\alpha/2} \sqrt{\dfrac{\sigma_1^2}{n_1} + \dfrac{\sigma_2^2}{n_2}}$. Estimating σ_1^2 and σ_2^2 with s_1^2 and s_2^2, the approximate 95% confidence interval is

$$(12.7 - 7.4) \pm 1.96 \sqrt{\dfrac{1.38}{35} + \dfrac{4.14}{49}} = 5.3 \pm .690 \quad or \quad 4.61 < \mu_1 - \mu_2 < 5.99 \ .$$

b Since the value $\mu_1 - \mu_2 = 0$ is not in the confidence interval, it is not likely that $\mu_1 = \mu_2$. You should conclude that there is a difference in the two population means.

8.43 **a** The 99% confidence interval for $\mu_1 - \mu_2$ is approximately

$$(\overline{x}_1 - \overline{x}_2) \pm 2.58 \sqrt{\dfrac{s_1^2}{n_1} + \dfrac{s_2^2}{n_2}}$$

$$(125.2 - 123.7) \pm 2.58 \sqrt{\dfrac{5.6^2}{100} + \dfrac{6.8^2}{100}}$$

$$1.5 \pm 2.27 \quad or \quad -.77 < (\mu_1 - \mu_2) < 3.77$$

b Since the value $\mu_1 - \mu_2 = 0$ is in the confidence interval, it is possible that $\mu_1 = \mu_2$. You should not conclude that there is a difference in the two population means.

8.45 The following information is available:

$$n_1 = n_2 = 30 \qquad \overline{x}_1 = 167.1 \qquad \overline{x}_2 = 140.9$$
$$s_1 = 24.3 \qquad s_2 = 17.6$$

The 95% confidence interval for $\mu_1 - \mu_2$ is approximately

$$(\overline{x}_1 - \overline{x}_2) \pm 1.96 \sqrt{\dfrac{s_1^2}{n_1} + \dfrac{s_2^2}{n_2}}$$

$$(167.1 - 140.9) \pm 1.96 \sqrt{\dfrac{(24.3)^2}{30} + \dfrac{(17.6)^2}{30}}$$

$$26.2 \pm 10.737 \quad or \quad 15.463 < (\mu_1 - \mu_2) < 36.937$$

In repeated sampling, 95% of all intervals constructed in this manner will enclose $\mu_1 - \mu_2$. Hence, we are fairly certain that this particular interval contains $\mu_1 - \mu_2$.

8.47 **a** The parameter to be estimated is μ, the mean score for the posttest for all BACC classes. The 95% confidence interval is approximately

$$\overline{x} \pm 1.96 \dfrac{s}{\sqrt{n}} = 18.5 \pm 1.96 \dfrac{8.03}{\sqrt{365}} = 18.5 \pm .824 \text{ or } 17.676 < \mu < 19.324$$

b The parameter to be estimated is μ, the mean score for the posttest for all traditional classes. The 95% confidence interval is approximately

$$\overline{x} \pm 1.96 \dfrac{s}{\sqrt{n}} = 16.5 \pm 1.96 \dfrac{6.96}{\sqrt{298}} = 16.5 \pm .790 \text{ or } 15.710 < \mu < 17.290$$

c Now we are interested in the difference between posttest means, $\mu_1 - \mu_2$, for BACC versus traditional classes. The 95% confidence interval for $\mu_1 - \mu_2$ is approximately

$$\left(\bar{x}_1 - \bar{x}_2\right) \pm 1.96\sqrt{\frac{s_1^2}{n_1} + \frac{s_2^2}{n_2}}$$

$$\left(18.5 - 16.5\right) \pm 1.96\sqrt{\frac{8.03^2}{365} + \frac{6.96^2}{298}}$$

$$2.0 \pm 1.142 \quad \text{or} \quad .858 < \left(\mu_1 - \mu_2\right) < 3.142$$

d Since the confidence interval in part **c** has two positive endpoints, it does not contain the value $\mu_1 - \mu_2 = 0$. Hence, it is not likely that the means are equal. It appears that there is a real difference in the mean scores.

8.49   **a**   The point estimate of the difference $\mu_1 - \mu_2$ is

$$\bar{x}_1 - \bar{x}_2 = 56,202 - 50,657 = 5545$$

and the margin of error is

$$1.96\sqrt{\frac{\sigma_1^2}{n_1} + \frac{\sigma_2^2}{n_2}} \approx 1.96\sqrt{\frac{2225^2}{50} + \frac{2375^2}{50}} = 902.08$$

b   Since the margin of error does not allow the estimate of the difference $\mu_1 - \mu_2$ to be negative—the lower limit is $5545 - 902.08 = 4642.92$—it is likely that the mean for engineering majors is larger than the mean for computer science majors.

8.51   Refer to Exercise 8.20.
   a   The 95% confidence interval for $\mu_1 - \mu_2$ is approximately

$$\left(\bar{x}_1 - \bar{x}_2\right) \pm 1.96\sqrt{\frac{s_1^2}{n_1} + \frac{s_2^2}{n_2}}$$

$$\left(150 - 165\right) \pm 1.96\sqrt{\frac{17.2^2}{50} + \frac{22.5^2}{50}}$$

$$-15 \pm 7.85 \quad \text{or} \quad -22.85 < \left(\mu_1 - \mu_2\right) < -7.15$$

   b   The 99% confidence interval for $\mu_1 - \mu_2$ is approximately

$$\left(\bar{x}_1 - \bar{x}_2\right) \pm 2.58\sqrt{\frac{s_1^2}{n_1} + \frac{s_2^2}{n_2}}$$

$$\left(165 - 125\right) \pm 2.58\sqrt{\frac{22.5^2}{50} + \frac{12.8^2}{50}}$$

$$40 \pm 9.44 \quad \text{or} \quad 30.56 < \left(\mu_1 - \mu_2\right) < 49.44$$

   c   Neither of the intervals contain the value $\left(\mu_1 - \mu_2\right) = 0$. If $\left(\mu_1 - \mu_2\right) = 0$ is contained in the confidence interval, then it is not unlikely that μ_1 could equal μ_2, implying no difference in the average room rates for the two hotels. This would be of interest to the experimenter.
   d   Since neither confidence interval contains the value $\mu_1 - \mu_2 = 0$, it is not likely that the means are equal. You should conclude that there is a difference in the average room rates for the Marriott and Westin and also for the Westin and the Doubletree chains.

8.53 The 95% confidence interval for $\mu_1 - \mu_2$ is approximately

$$\left(\bar{x}_1 - \bar{x}_2\right) \pm 1.96\sqrt{\frac{s_1^2}{n_1} + \frac{s_2^2}{n_2}}$$

$$(98.11 - 98.39) \pm 1.96\sqrt{\frac{.7^2}{65} + \frac{.74^2}{65}}$$

$$-.28 \pm .248 \quad \text{or} \quad -.528 < \left(\mu_1 - \mu_2\right) < -.032$$

b Since the confidence interval in part **a** has two negative endpoints, it does not contain the value $\mu_1 - \mu_2 = 0$. Hence, it is not likely that the means are equal. It appears that there is a real difference in the mean temperatures for males and females.

8.55 **a** Calculate $\hat{p}_1 = \dfrac{x_1}{n_1} = \dfrac{337}{800} = .42$ and $\hat{p}_2 = \dfrac{x_2}{n_2} = \dfrac{374}{640} = .58$. The approximate 90% confidence interval is

$$\left(\hat{p}_1 - \hat{p}_2\right) \pm 1.645\sqrt{\frac{\hat{p}_1\hat{q}_1}{n_1} + \frac{\hat{p}_2\hat{q}_2}{n_2}}$$

$$(.42 - .58) \pm 1.645\sqrt{\frac{.42(.58)}{800} + \frac{.58(.42)}{640}}$$

$$-.16 \pm .043 \quad \text{or} \quad -.203 < \left(p_1 - p_2\right) < -.117$$

b The two binomial samples must be random and independent and the sample sizes must be large enough that the distributions of $\hat{p}_1$ and $\hat{p}_2$ are approximately normal. Assuming that the samples are random, these conditions are met in this exercise.

8.57 **a** Calculate $\hat{p}_1 = \dfrac{x_1}{n_1} = \dfrac{12}{56} = .214$ and $\hat{p}_2 = \dfrac{x_2}{n_2} = \dfrac{8}{32} = .25$. The approximate 95% confidence interval is

$$\left(\hat{p}_1 - \hat{p}_2\right) \pm 1.96\sqrt{\frac{\hat{p}_1\hat{q}_1}{n_1} + \frac{\hat{p}_2\hat{q}_2}{n_2}}$$

$$(.214 - .25) \pm 1.96\sqrt{\frac{.214(.786)}{56} + \frac{.25(.75)}{32}}$$

$$-.036 \pm .185 \quad \text{or} \quad -.221 < \left(p_1 - p_2\right) < .149$$

b Since the value $p_1 - p_2 = 0$ is in the confidence interval, it is possible that $p_1 = p_2$. You should not conclude that there is a difference in the proportion of red candies in plain and peanut M&Ms.

8.59 **a** With $\hat{p}_1 = \dfrac{x_1}{1001} = .45$ and $\hat{p}_2 = \dfrac{x_2}{1001} = .51$. The approximate 99% confidence interval is

$$\left(\hat{p}_1 - \hat{p}_2\right) \pm 2.58\sqrt{\frac{\hat{p}_1\hat{q}_1}{n_1} + \frac{\hat{p}_2\hat{q}_2}{n_2}}$$

$$(.45 - .51) \pm 2.58\sqrt{\frac{.45(.55)}{1001} + \frac{.51(.49)}{1001}}$$

$$-.06 \pm .058 \quad \text{or} \quad -.118 < \left(p_1 - p_2\right) < -.002$$

b Since the interval in part **a** contains only negative values of $p_1 - p_2$, it is likely that $p_1 - p_2 < 0 \Rightarrow p_1 < p_2$. This would indicate that the proportion of adults who claim to be fans is higher in November than in March.

8.61   **a**   With $\hat{p}_1 = \dfrac{x_1}{96} = .62$ and $\hat{p}_2 = \dfrac{x_2}{105} = .35$. The approximate 99% confidence interval is

$$(\hat{p}_1 - \hat{p}_2) \pm 2.58 \sqrt{\frac{\hat{p}_1 \hat{q}_1}{n_1} + \frac{\hat{p}_2 \hat{q}_2}{n_2}}$$

$$(.62 - .35) \pm 2.58 \sqrt{\frac{.62(.38)}{96} + \frac{.35(.65)}{105}}$$

$$.27 \pm .175 \quad \text{or} \quad .095 < (p_1 - p_2) < .445$$

b   Since the value $p_1 - p_2 = 0$ is not in the confidence interval, it is not likely that $p_1 = p_2$. You should conclude that there is a difference in the proportion of people with colds for the two groups.

c   The data shows the opposite effect from what you might expect. It appears that coming into contact with more people actually *decreases* the probability of getting a cold. Perhaps the more you are out among a variety of people, the more immune your system becomes to cold germs and viruses.

8.63   Calculate $\hat{p}_1 = \dfrac{x_1}{n_1} = \dfrac{120}{180} = .7$ and $\hat{p}_2 = \dfrac{x_2}{n_2} = \dfrac{54}{100} = .54$. The approximate 90% confidence interval is

$$(\hat{p}_1 - \hat{p}_2) \pm 1.645 \sqrt{\frac{\hat{p}_1 \hat{q}_1}{n_1} + \frac{\hat{p}_2 \hat{q}_2}{n_2}}$$

$$(.7 - .54) \pm 1.645 \sqrt{\frac{.7(.3)}{180} + \frac{.54(.46)}{100}}$$

$$.16 \pm .099 \quad \text{or} \quad .061 < (p_1 - p_2) < .259$$

Intervals constructed in this manner will enclose the true value of $p_1 - p_2$ 95% of the time in repeated sampling. Hence, we are fairly certain that this particular interval encloses $p_1 - p_2$.

8.65   **a**   Calculate $\hat{p}_1 = \dfrac{x_1}{200} = .93$ and $\hat{p}_2 = \dfrac{x_2}{450} = .96$. The approximate 99% confidence interval is

$$(\hat{p}_1 - \hat{p}_2) \pm 2.58 \sqrt{\frac{\hat{p}_1 \hat{q}_1}{n_1} + \frac{\hat{p}_2 \hat{q}_2}{n_2}}$$

$$(.93 - .96) \pm 2.58 \sqrt{\frac{.93(.07)}{200} + \frac{.96(.04)}{450}}$$

$$-.03 \pm .052 \quad \text{or} \quad -.082 < (p_1 - p_2) < .022$$

b   Since the value $p_1 - p_2 = 0$ is in the confidence interval, it is possible that $p_1 = p_2$. You should not conclude that there is a difference in the proportion of people who experience pain relief when using one pain reliever or the other.

8.67   The parameter to be estimated is the population mean μ and the 90% upper confidence bound is calculated using a value $z_\alpha = z_{.10} = 1.28$.

a   The upper bound is approximately

$$\bar{x} + 1.28 \frac{s}{\sqrt{n}} = 75 + 1.28 \sqrt{\frac{65}{40}} = 75 + 1.63 \quad \text{or} \quad \mu < 76.63$$

b   The upper bound is approximately

$$\bar{x} + 1.28 \frac{s}{\sqrt{n}} = 1.6 + 1.28 \frac{2.3}{\sqrt{100}} = 1.6 + .29 \quad \text{or} \quad \mu < 1.89$$

8.69   For the difference $\mu_1 - \mu_2$ in the population means for two quantitative populations, the 95% upper confidence bound uses $z_{.05} = 1.645$ and is calculated as

$$(\bar{x}_1 - \bar{x}_2) + 1.645\sqrt{\frac{s_1^2}{n_1} + \frac{s_2^2}{n_2}} = (12 - 10) + 1.645\sqrt{\frac{5^2}{50} + \frac{7^2}{50}}.$$

$$2 + 2.00 \quad \text{or} \quad (\mu_1 - \mu_2) < 4$$

8.71 It is necessary to find the sample size required to estimate a certain parameter to within a given bound with confidence $(1-\alpha)$. Recall from Section 8.5 that we may estimate a parameter with $(1-\alpha)$ confidence within the interval (estimator) $\pm z_{\alpha/2} \times$ (std error of estimator). Thus, $z_{\alpha/2} \times$ (std error of estimator) provides the margin of error with $(1-\alpha)$ confidence. The experimenter will specify a given bound B. If we let $z_{\alpha/2} \times$ (std error of estimator) $\leq B$, we will be $(1-\alpha)$ confident that the estimator will lie within B units of the parameter of interest. For this exercise, B = .04 for the binomial estimator $\hat{p}$, where $SE(\hat{p}) = \sqrt{\frac{pq}{n}}$. Assuming maximum variation, which occurs if $p = .3$ (since we suspect that $.1 < p < .3$) and $z_{.025} = 1.96$, we have

$$1.96\sigma_{\hat{p}} \leq B \Rightarrow 1.96\sqrt{\frac{pq}{n}} \leq B$$

$$1.96\sqrt{\frac{.3(.7)}{n}} \leq .04 \Rightarrow \sqrt{n} \geq \frac{1.96\sqrt{.3(.7)}}{.04} \Rightarrow n \geq 504.21 \quad \text{or} \quad n \geq 505$$

8.73 In this exercise, the parameter of interest is $p_1 - p_2$, $n_1 = n_2 = n$, and B = .05. Since we have no prior knowledge about p_1 and p_2, we assume the largest possible variation, which occurs if $p_1 = p_2 = .5$. Then

$$z_{\alpha/2} \times (\text{std error of } \hat{p}_1 - \hat{p}_2) \leq B$$

$$z_{.01}\sqrt{\frac{p_1 q_1}{n_1} + \frac{p_2 q_2}{n_2}} \leq .05 \Rightarrow 2.33\sqrt{\frac{(.5)(.5)}{n} + \frac{(.5)(.5)}{n}} \leq .05$$

$$\sqrt{n} \geq \frac{2.33\sqrt{.5}}{.05} \Rightarrow n \geq 1085.78 \quad \text{or} \quad n_1 = n_2 = 1086$$

8.75 **a** The sample should be selected randomly. Voter registration lists, DMV lists, telephone listings may provide possible lists from which you might choose. Make sure that your lists do not systematically exclude any segment of the population, which might bias your results.
b To estimate binomial proportions p in the survey, choose a common value of $p = .5$ to maximize the possible error. With B = .01, solve for n:

$$1.96\sqrt{\frac{pq}{n}} \leq .01 \Rightarrow 1.96\sqrt{\frac{.5(.5)}{n}} \leq .01 \Rightarrow n \geq 9604$$

8.77 In this exercise, the parameter of interest is $\mu_1 - \mu_2$, $n_1 = n_2 = n$, and $s \approx R/4 = 104/4 = 26$. Then we must have

$$z_{\alpha/2} \times (\text{std error of } \bar{x}_1 - \bar{x}_2) \leq B$$

$$2.58\sqrt{\frac{\sigma_1^2}{n_1} + \frac{\sigma_2^2}{n_2}} \leq 5 \Rightarrow 2.58\sqrt{\frac{26^2}{n} + \frac{26^2}{n}} \leq 5$$

$$\sqrt{n} \geq \frac{2.58\sqrt{1352}}{5} \Rightarrow n \geq 359.98 \quad \text{or} \quad n_1 = n_2 = 360$$

8.79   The margin of error in estimation has bound B = 2 days. Assuming $\sigma \approx 10$, we must have

$$1.96\frac{\sigma}{\sqrt{n}} \le 2 \quad \Rightarrow \quad 1.96\frac{10}{\sqrt{n}} \le 2$$

$$\sqrt{n} \ge \frac{1.96(10)}{2} = 9.8 \quad \Rightarrow \quad n \ge 96.04 \text{ or } n \ge 97$$

Therefore, we must include 97 hunters in the survey in order to estimate the mean number of days of hunting per hunter to within 2 days.

8.81   There are now two populations of interest and the parameter to be estimated is $\mu_1 - \mu_2$. The bound on the margin of error is B = .1, $n_1 = n_2 = n$, and $\sigma_1^2 \approx \sigma_2^2 \approx .25$. Then we must have

$$1.645\sqrt{\frac{\sigma_1^2}{n_1} + \frac{\sigma_2^2}{n_2}} \le .1 \quad \Rightarrow \quad 1.645\sqrt{\frac{.25}{n} + \frac{.25}{n}} \le .1$$

$$\sqrt{n} \ge \frac{1.645\sqrt{.5}}{.1} \quad \Rightarrow \quad n \ge 135.30$$

or $n_1 = n_2 = 136$ samples should be selected at each location.

8.83   It is given that $n_1 = n_2 = n$ and that B = 5. From Exercise 8.45, $s_1 = 24.3$ and $s_2 = 17.6$. Using these values to estimate σ_1 and σ_2, the following inequality must be solved:

$$1.645\sqrt{\frac{\sigma_1^2}{n_1} + \frac{\sigma_2^2}{n_2}} \le 5 \quad \Rightarrow \quad 1.645\sqrt{\frac{24.3^2}{n} + \frac{17.6^2}{n}} \le 5$$

$$n \ge 97.444 \quad \text{or} \quad n_1 = n_2 = 98$$

8.85   **a**   The point estimate of μ is $\bar{x} = 29.1$ and the margin of error in estimation with $s = 3.9$ and $n = 64$ is

$$1.96\sigma_{\bar{x}} = 1.96\frac{\sigma}{\sqrt{n}} \approx 1.96\frac{s}{\sqrt{n}} = 1.96\left(\frac{3.9}{\sqrt{64}}\right) = .9555$$

   b   The approximate 90% confidence interval is

$$\bar{x} \pm 1.645\frac{s}{\sqrt{n}} = 29.1 \pm 1.645\frac{3.9}{\sqrt{64}} = 29.1 \pm .802 \quad \text{or} \quad 28.298 < \mu < 29.902$$

Intervals constructed in this manner enclose the true value of μ 90% of the time in repeated sampling. Therefore, we are fairly certain that this particular interval encloses μ.

   c   The approximate 90% lower confidence bound is

$$\bar{x} - 1.28\frac{s}{\sqrt{n}} = 29.1 - 1.28\frac{3.9}{\sqrt{64}} = 28.48 \quad \text{or} \quad \mu > 28.48$$

   d   With B = .5, $\sigma \approx 3.9$, and $1 - \alpha = .95$, we must solve for n in the following inequality:

$$1.96\frac{\sigma}{\sqrt{n}} \le B \quad \Rightarrow \quad 1.96\frac{3.9}{\sqrt{n}} \le .5$$

$$\sqrt{n} \ge 15.288 \quad \Rightarrow \quad n \ge 233.723 \text{ or } n \ge 234$$

8.87   Refer to Exercise 8.86, with B = .2, $1 - \alpha = .95$, $n_1 = n_2 = n$, $s_1 = .8$ and $s_2 = 1.3$. Using these values to estimate σ_1 and σ_2, the following inequality must be solved:

$$1.96\sqrt{\frac{\sigma_1^2}{n_1} + \frac{\sigma_2^2}{n_2}} \le .2 \quad \Rightarrow \quad 1.96\sqrt{\frac{(.8)^2}{n} + \frac{(1.3)^2}{n}} \le .2$$

$$\sqrt{n} \ge 14.959 \quad \Rightarrow \quad n \ge 223.77 \text{ or } n_1 = n_2 = 224$$

8.89 Assuming maximum variation with $p = .5$, solve

$$1.645\sqrt{\frac{pq}{n}} \le .025$$

$$\sqrt{n} \ge \frac{1.645\sqrt{.5(.5)}}{.025} = 32.9 \Rightarrow n \ge 1082.41 \quad \text{or} \quad n \ge 1083$$

8.91 Assuming maximum variation $(p_1 = p_2 = .5)$ and $n_1 = n_2 = n$, the inequality to be solved is

$$z_{.005}\sqrt{\frac{p_1 q_1}{n_1} + \frac{p_2 q_2}{n_2}} \le .06 \Rightarrow 2.58\sqrt{\frac{(.5)(.5)}{n} + \frac{(.5)(.5)}{n}} \le .06$$

$$\sqrt{n} \ge 30.406 \Rightarrow n \ge 924.5 \quad \text{or} \quad n_1 = n_2 = 925$$

8.93 **a** Define sample #1 as the responses of the 72 people under the age of 34 and sample #2 as the responses of the 55 people who are 65 or older. Then $\hat{p}_1 = .37$ and $\hat{p}_2 = .13$.

b The approximate 95% confidence interval is

$$(\hat{p}_1 - \hat{p}_2) \pm 1.96\sqrt{\frac{\hat{p}_1 \hat{q}_1}{n_1} + \frac{\hat{p}_2 \hat{q}_2}{n_2}}$$

$$(.37 - .13) \pm 1.96\sqrt{\frac{.37(.63)}{72} + \frac{.13(.87)}{55}}$$

$$.24 \pm .143 \quad \text{or} \quad .097 < (p_1 - p_2) < .383$$

c Since the value $p_1 - p_2 = 0$ is not in the confidence interval, it is unlikely that $p_1 = p_2$. You should conclude that there is a difference in the proportion of people in the two groups who more likely to "haggle". In fact, since all the probable values of $p_1 - p_2$ are positive, the proportion of young people who "haggle" appears to be larger than the proportion of older people.

8.95 The approximate 90% confidence interval for μ is

$$\bar{x} \pm 1.645\frac{s}{\sqrt{n}} = 9.7 \pm 1.645\frac{5.8}{\sqrt{35}} = 9.7 \pm 1.613 \text{ or } 8.087 < \mu < 11.313$$

8.97 Assume that $\sigma = 2.5$ and the desired bound is .5. Then

$$1.96\frac{\sigma}{\sqrt{n}} \le B \Rightarrow 1.96\frac{2.5}{\sqrt{n}} \le .5 \Rightarrow n \ge 96.04 \quad \text{or} \quad n \ge 97$$

8.99 The approximate 95% confidence interval for μ is

$$\bar{x} \pm 1.96\frac{s}{\sqrt{n}} = 34 \pm 1.96\frac{3}{\sqrt{100}} = 34 \pm .59 \text{ or } 33.41 < \mu < 34.59$$

8.101 **a** There are problems with non-response and households without "landlines".

b If you use $p = .5$ as an estimate for p, the margin of error is approximately

$$\pm 1.96\sqrt{\frac{.5(.5)}{2250}} = \pm .021$$

c To reduce the margin of error in part **b** to $\pm .01$, solve for n in the equation

$$1.96\sqrt{\frac{.5(.5)}{n}} = .01 \Rightarrow \sqrt{n} = \frac{1.96(.5)}{.01} = 98 \Rightarrow n = 9604$$

8.103 **a** Answers will vary. We would suspect that there will be a few tickets with a very high price tag, in which case the distribution of prices may be skewed to the right.

b Since the sample size n is large, the Central Limit Theorem insures the approximate normality of the sample mean $\bar{x}$, used in this chapter to estimate the population mean μ.

c A 95% confidence interval for μ, the mean price of a ticket, is approximated as

$$82.50 \pm 1.96 \frac{75.25}{\sqrt{50}} = 82.50 \pm 20.86 \text{ or } 61.64 < \mu < 103.36$$

The value $\mu = 75.50$ is one of the possible values given in the confidence interval. Hence, there is no reason to question the claimed average.

8.105 It is assumed that $p = .2$ and that the desired bound is .01. Hence,

$$1.96\sqrt{\frac{pq}{n}} \leq .01 \implies \sqrt{n} \geq \frac{1.96\sqrt{.05(.95)}}{.01} = 42.72$$

$$n \geq 1824.76 \text{ or } n \geq 1825$$

8.107 Ten samples of $n = 400$ printed circuit boards were tested and a $100(1-\alpha)\%$ confidence interval for p was constructed for each of the ten samples. For this exercise, $100(1-\alpha)\% = 90\%$, or $\alpha = .10$. The object is to find the probability that exactly one of the intervals will not enclose the true value of p. Hence, the situation descries a binomial experiment with $n = 10$ and

$$p^* = P[\text{an interval will not contain the true value of } p]$$
$$x = \text{number of intervals which do not enclose } p$$

By definition of a 90% confidence interval, it can be said that 90% of the intervals generated in repeated sampling will contain the true value of p; 10% of the intervals will not contain p. Thus, $p^* = .1$ and the desired probabilities are calculated using the methods of Chapter 4.

1 $P[\text{exactly one of the intervals fails to contain } p] = P[x = 1] = C_1^{10}(.1)^1(.9)^9 = .3874$

2 $P[\text{at least one}] = 1 - P[x = 0] = 1 - C_0^{10}(.1)^0(.9)^{10} = 1 - .349 = .651$

8.109 **a** The approximate 95% confidence interval for μ is

$$\bar{x} \pm 1.96 \frac{s}{\sqrt{n}} = 2.962 \pm 1.96 \frac{.529}{\sqrt{69}} = 2.962 \pm .125$$

or $2.837 < \mu < 3.087$.

b In order to cut the interval in half, the sample size must increase by 4. If this is done, the new half-width of the confidence interval is

$$1.96 \frac{\sigma}{\sqrt{4n}} = \frac{1}{2}\left(1.96 \frac{\sigma}{\sqrt{n}}\right).$$

Hence, in this case, the new sample size is $4(69) = 276$.

8.111 The approximate 95% confidence interval for μ is

$$\bar{x} \pm 1.96 \frac{s}{\sqrt{n}} = 13.45 \pm 1.96 \frac{2.84}{\sqrt{36}} = 13.45 \pm .93$$

or $12.52 < \mu < 14.38$. Since $\mu = 16.92$ is not in this interval, it is unlikely that the average paid to bus drivers in Auburn, Washington is the same as the average paid by other school districts. Since the possible values given in the confidence interval are all lower than $16.92, we can conclude that Auburn, Washington pays significantly more per hour than other school districts.

8.113   The approximate 98% confidence interval for μ is

$$\bar{x} \pm 2.33 \frac{s}{\sqrt{n}} = 2.705 \pm 2.33 \frac{.028}{\sqrt{36}} = 2.705 \pm .011$$

or $2.694 < \mu < 2.716$.

8.115   Calculate $\hat{p} = \dfrac{80}{400} = .20$. Then the approximate 95% confidence interval for p is

$$\hat{p} \pm 1.96 \sqrt{\frac{\hat{p}\hat{q}}{n}} = .20 \pm 1.96 \sqrt{\frac{.20(.80)}{400}} = .20 \pm .039$$

or $.161 < p < .239$.

8.117   For this exercise, $B = .08$ for the binomial estimator $\hat{p}$, where $SE(\hat{p}) = \sqrt{\dfrac{pq}{n}}$. If $p = .3$, we have

$$1.96 \sqrt{\frac{pq}{n}} \leq B \Rightarrow 1.96 \sqrt{\frac{.2(.8)}{n}} \leq .08$$

$$\Rightarrow \sqrt{n} \geq \frac{1.96\sqrt{.2(.8)}}{.08} \Rightarrow n \geq 9.8 \quad \text{or} \quad n \geq 96.04$$

or $n \geq 97$.

9: Large-Sample Tests of Hypotheses

9.1   **a**   The critical value that separates the rejection and nonrejection regions for a right-tailed test based on a z-statistic will be a value of z (called z_α) such that $P(z > z_\alpha) = \alpha = .01$. That is, $z_{.01} = 2.33$ (see the figure below). The null hypothesis H_0 will be rejected if $z > 2.33$.

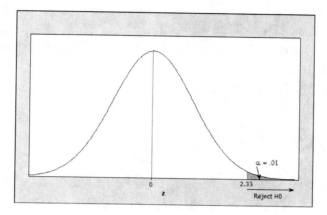

   b   For a two-tailed test with $\alpha = .05$, the critical value for the rejection region cuts off $\alpha/2 = .025$ in the two tails of the z distribution in the figure below, so that $z_{.025} = 1.96$. The null hypothesis H_0 will be rejected if $z > 1.96$ or $z < -1.96$ (which you can also write as $|z| > 1.96$).

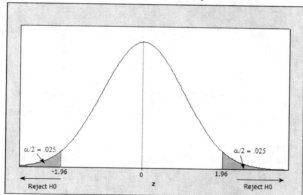

9.3   **a**   The critical value that separates the rejection and nonrejection regions for a lower-tailed test based on a z-statistic will be a value of z that cuts off $\alpha = .01$ in the left tail of the distribution. That is, $z_{.01} = -2.33$. This rejection region is in the lower tail of the z distribution. The null hypothesis H_0 will be rejected if $z < -2.33$.

   b   For a two-tailed test with $\alpha = .01$, the critical value for the rejection region cuts off $\alpha/2 = .005$ in each of the two tails of the z distribution, so that $z_{.005} = 2.58$. The null hypothesis H_0 will be rejected if $z > 2.58$ or $z < -2.58$ (which you can also write as $|z| > 2.58$).

   c   If the observed value of the test statistic is $z = -2.41$, the null hypothesis H_0 will be rejected at the 1% level for the rejection region in part (a). For the rejection region in part (b), we do not reject H_0.

9.5   Use the guidelines for statistical significance in Section 9.3. The smaller the p-value, the more evidence there is in favor of rejecting H_0. For part **a**, p-value $= .1251$ is not statistically significant; H_0 is not rejected. For part **b**, p-value $= .0054$ is less than $.01$ and the results are highly significant; H_0 should be rejected. For part **c,** p-value $= .0351$ is between $.01$ and $.05$. The results are significant at the 5% level, but not at the 1% level ($P < .05$).

9.7 **a** Since this is a right-tailed test, the p-value is the area under the standard normal distribution to the right of $z = 2.04$:
$$p\text{-value} = P(z > 2.04) = 1 - .9793 = .0207$$

b The p-value, $.0207$, is less than $\alpha = .05$, and the null hypothesis is rejected at the 5% level of significance. There is sufficient evidence to indicate that $\mu > 2.3$.

c The conclusions reached using the **critical value approach** and the **p-value approach** are identical.

9.9 The hypotheses to be tested are
$$H_0 : \mu = 28 \quad \text{versus} \quad H_a : \mu \neq 28$$
and the test statistic is
$$z = \frac{\bar{x} - \mu_0}{\sigma/\sqrt{n}} \approx \frac{\bar{x} - \mu_0}{s/\sqrt{n}} = \frac{26.8 - 28}{6.5/\sqrt{100}} = -1.85$$

with p-value $= P(|z| > 1.85) = 2(.0322) = .0644$. To draw a conclusion from the p-value, use the guidelines for statistical significance in Section 9.3. Since the p-value is greater than $.05$, the null hypothesis should not be rejected. There is insufficient evidence to indicate that the mean is different from 28. (Some researchers might report these results as *tending towards significance*.)

9.11 **a** In order to make sure that the average weight was one pound, you would test
$$H_0 : \mu = 1 \quad \text{versus} \quad H_a : \mu \neq 1$$

b-c The test statistic is
$$z = \frac{\bar{x} - \mu_0}{\sigma/\sqrt{n}} \approx \frac{\bar{x} - \mu_0}{s/\sqrt{n}} = \frac{1.01 - 1}{.18/\sqrt{35}} = .33$$

with p-value $= P(|z| > .33) = 2(.3707) = .7414$. Since the p-value is greater than $.05$, the null hypothesis should not be rejected. The manager should report that there is insufficient evidence to indicate that the mean is different from 1.

9.13 **a-b** We want to test the null hypothesis that μ is, in fact, 80% against the alternative that it is not:
$$H_0 : \mu = 80 \quad \text{versus} \quad H_a : \mu \neq 80$$
Since the exercise does not specify $\mu < 80$ or $\mu > 80$, we are interested in a two directional alternative, $\mu \neq 80$.

c The test statistic is
$$z = \frac{\bar{x} - \mu_0}{\sigma/\sqrt{n}} \approx \frac{\bar{x} - \mu_0}{s/\sqrt{n}} = \frac{79.7 - 80}{.8/\sqrt{100}} = -3.75$$

The rejection region with $\alpha = .05$ is determined by a critical value of z such that
$$P(z < -z_0) + P(z > z_0) = \frac{\alpha}{2} + \frac{\alpha}{2} = .05$$

This value is $z_0 = 1.96$ (see the figure in Exercise 9.1**b**). Hence, H_0 will be rejected if $z > 1.96$ or $z < -1.96$. The observed value, $z = -3.75$, falls in the rejection region and H_0 is rejected. There is sufficient evidence to refute the manufacturer's claim. The probability that we have made an incorrect decision is $\alpha = .05$.

9.15   **a** The hypothesis to be tested is

$$H_0 : \mu = 7.4 \quad \text{versus} \quad H_a : \mu > 7.4$$

and the test statistic is

$$z = \frac{\overline{x} - \mu_0}{\sigma/\sqrt{n}} \approx \frac{\overline{x} - \mu_0}{s/\sqrt{n}} = \frac{7.9 - 7.4}{1.9/\sqrt{100}} = 2.63$$

with p-value $= P(z > 2.63) = 1 - .9957 = .0043$. To draw a conclusion from the p-value, use the guidelines for statistical significance in Section 9.3. Since the p-value is less than .01, the test results are highly significant. We can reject H_0 at both the 1% and 5% levels of significance.
b You could claim that you work significantly fewer hours than those without a college education.
c If you were not a college graduate, you might just report that you work an average of more than 7.4 hours per week..

9.17   The hypothesis to be tested is

$$H_0 : \mu = 5.97 \quad \text{versus} \quad H_a : \mu > 5.97$$

and the test statistic is

$$z = \frac{\overline{x} - \mu_0}{\sigma/\sqrt{n}} \approx \frac{\overline{x} - \mu_0}{s/\sqrt{n}} = \frac{9.8 - 5.97}{1.95/\sqrt{31}} = 10.94$$

with p-value $= P(z > 10.94) < 1 - .9998 = .0002$ (or p-value ≈ 0). Since the p-value is less than .05, the null hypothesis is rejected. There is sufficient evidence to indicate that the average diameter of the tendon for patients with AT is greater than 5.97 mm.

9.19   The hypothesis of interest is one-tailed:

$$H_0 : \mu_1 - \mu_2 = 0 \quad \text{versus} \quad H_a : \mu_1 - \mu_2 < 0$$

The test statistic, calculated under the assumption that $\mu_1 - \mu_2 = 0$, is

$$z \approx \frac{(\overline{x}_1 - \overline{x}_2) - 0}{\sqrt{\dfrac{s_1^2}{n_1} + \dfrac{s_2^2}{n_2}}} = \frac{1.24 - 1.31}{\sqrt{\dfrac{.056}{36} + \dfrac{.054}{45}}} = -1.334$$

with the unknown σ_1^2 and σ_2^2 estimated by s_1^2 and s_2^2, respectively. The student can use one of two methods for decision making.
_p_-value approach: Calculate p-value $= P(z < -1.33) = .0918$. Since this p-value is greater than .05, the null hypothesis is not rejected. There is insufficient evidence to indicate that the mean for population 1 is smaller than the mean for population 2.
Critical value approach: The rejection region with $\alpha = .05$, is $z < -1.645$. Since the observed value of z does not fall in the rejection region, H_0 is not rejected. There is insufficient evidence to indicate that the mean for population 1 is smaller than the mean for population 2.

9.21   **a** The hypothesis of interest is one-tailed:

$$H_0 : \mu_1 - \mu_2 = 0 \quad \text{versus} \quad H_a : \mu_1 - \mu_2 > 0$$

b The test statistic, calculated under the assumption that $\mu_1 - \mu_2 = 0$, is

$$z \approx \frac{(\overline{x}_1 - \overline{x}_2) - 0}{\sqrt{\dfrac{s_1^2}{n_1} + \dfrac{s_2^2}{n_2}}} = \frac{6.9 - 5.8}{\sqrt{\dfrac{(2.9)^2}{35} + \dfrac{(1.2)^2}{35}}} = 2.074$$

The rejection region with $\alpha = .05$, is $z > 1.645$ and H_0 is rejected. There is evidence to indicate that $\mu_1 - \mu_2 > 0$, or $\mu_1 > \mu_2$. That is, there is reason to believe that Vitamin C reduces the mean time to recover.

105

9.23   **a**   The hypothesis of interest is two-tailed:
$$H_0 : \mu_1 - \mu_2 = 0 \quad \text{versus} \quad H_a : \mu_1 - \mu_2 \neq 0$$

The test statistic, calculated under the assumption that $\mu_1 - \mu_2 = 0$, is
$$z \approx \frac{(\bar{x}_1 - \bar{x}_2) - 0}{\sqrt{\frac{s_1^2}{n_1} + \frac{s_2^2}{n_2}}} = \frac{34.1 - 36}{\sqrt{\frac{(5.9)^2}{100} + \frac{(6.0)^2}{100}}} = -2.26$$

with p-value $= P(|z| > 2.26) = 2(.0119) = .0238$. Since the p-value is less than .05, the null hypothesis is rejected. There is evidence to indicate a difference in the mean lead levels for the two sections of the city.

b   From Section 8.6, the 95% confidence interval for $\mu_1 - \mu_2$ is approximately
$$(\bar{x}_1 - \bar{x}_2) \pm 1.96 \sqrt{\frac{s_1^2}{n_1} + \frac{s_2^2}{n_2}}$$
$$(34.1 - 36) \pm 1.96 \sqrt{\frac{5.9^2}{100} + \frac{6.0^2}{100}}$$
$$-1.9 \pm 1.65 \quad \text{or} \quad -3.55 < (\mu_1 - \mu_2) < -.25$$

c   Since the value $\mu_1 - \mu_2 = 5$ or $\mu_1 - \mu_2 = -5$ is not in the confidence interval in part **b**, it is not likely that the difference will be more than 5 ppm, and hence the statistical significance of the difference is not of practical importance to the engineers.

9.25   **a**   Most people would have no preconceived idea about which of the two hotels would have higher average room rates, and a two-tailed hypothesis would be appropriate:
$$H_0 : \mu_1 - \mu_2 = 0 \quad \text{versus} \quad H_a : \mu_1 - \mu_2 \neq 0$$

b   The test statistic is
$$z \approx \frac{(\bar{x}_1 - \bar{x}_2) - 0}{\sqrt{\frac{s_1^2}{n_1} + \frac{s_2^2}{n_2}}} = \frac{150 - 165}{\sqrt{\frac{17.2^2}{50} + \frac{22.5^2}{50}}} = -3.75$$

The rejection region, with $\alpha = .01$, is $|z| > 2.58$ and H_0 is rejected. There is evidence to indicate that there is a difference in the average room rates for the Marriott and the Westin hotels.

c   The p-value for this two-tailed test is
$$p\text{-value} = P(z > 3.75) + P(z < -3.75) \approx .0000$$

Since the p-value is less than $\alpha = .01$, the null hypothesis can be rejected at the 1% level. There is sufficient evidence to conclude that $\mu_1 - \mu_2 \neq 0$.

9.27   **a**   The hypothesis of interest is two-tailed:
$$H_0 : \mu_1 - \mu_2 = 0 \quad \text{versus} \quad H_a : \mu_1 - \mu_2 \neq 0$$

and the test statistic is
$$z \approx \frac{(\bar{x}_1 - \bar{x}_2) - 0}{\sqrt{\frac{s_1^2}{n_1} + \frac{s_2^2}{n_2}}} = \frac{349 - 287}{\sqrt{\frac{200^2}{1000} + \frac{200^2}{1000}}} = 6.93$$

with p-value $= P(|z| > 6.93) \approx 2(.0000) = .0000$. Since the p-value is less than .05, the null hypothesis is rejected. There is evidence to indicate a difference in the mean ticket prices for these two airports.

b   The 95% confidence interval for $\mu_1 - \mu_2$ is approximately

106

$$(\bar{x}_1 - \bar{x}_2) \pm 1.96 \sqrt{\frac{s_1^2}{n_1} + \frac{s_2^2}{n_2}}$$

$$(349 - 287) \pm 1.96 \sqrt{\frac{200^2}{1000} + \frac{200^2}{1000}}$$

$$62 \pm 17.53 \quad \text{or} \quad 44.47 < (\mu_1 - \mu_2) < 79.53$$

Since the value $\mu_1 - \mu_2 = 0$ does not fall in the interval in part **b**, it is not likely that $\mu_1 = \mu_2$. There is evidence to indicate that the means are different, confirming the conclusion in part **a**.

9.29 **a** The hypothesis of interest is two-tailed:

$$H_0 : \mu_1 - \mu_2 = 0 \quad \text{versus} \quad H_a : \mu_1 - \mu_2 \neq 0$$

and the test statistic is

$$z \approx \frac{(\bar{x}_1 - \bar{x}_2) - 0}{\sqrt{\frac{s_1^2}{n_1} + \frac{s_2^2}{n_2}}} = \frac{98.11 - 98.39}{\sqrt{\frac{.7^2}{65} + \frac{.74^2}{65}}} = -2.22$$

with p-value $= P\left(|z| > 2.22\right) = 2(1 - .9868) = .0264$. Since the p-value is between .01 and .05, the null hypothesis is rejected, and the results are significant. There is evidence to indicate a difference in the mean temperatures for men versus women.

b Since the p-value $= .0264$, we can reject H_0 at the 5% level (p-value $< .05$), but not at the 1% level (p-value $> .01$). Using the guidelines for significance given in Section 9.3 of the text, we declare the results statistically *significant*, but not *highly significant*.

9.31 **a** The hypothesis of interest is two-tailed:

$$H_0 : p = .4 \quad \text{versus} \quad H_a : p \neq .4$$

b-c It is given that $x = 529$ and $n = 1400$, so that $\hat{p} = \frac{x}{n} = \frac{529}{1400} = .378$. The test statistic is

$$z = \frac{\hat{p} - p_0}{\sqrt{\frac{p_0 q_0}{n}}} = \frac{.378 - .4}{\sqrt{\frac{.4(.6)}{1400}}} = -1.68$$

with p-value $= P\left(|z| > 1.68\right) = 2(.0465) = .093$. Since the p-value is not less than $\alpha = .01$, the null hypothesis is not rejected. There is insufficient evidence to indicate that p differs from .4.

9.33 **a** The hypothesis to be tested involves the binomial parameter p:

$$H_0 : p = .15 \quad \text{versus} \quad H_a : p < .15$$

where p is the proportion of parents who describe their children as overweight. For this test, $x = 68$ and $n = 750$, so that $\hat{p} = \frac{x}{n} = \frac{68}{750} = .091$, the test statistic is

$$z = \frac{\hat{p} - p_0}{\sqrt{\frac{p_0 q_0}{n}}} = \frac{.091 - .15}{\sqrt{\frac{.15(.85)}{750}}} = -4.53$$

b The rejection region is one-tailed, with $z < -1.645$ for $\alpha = .05$. Since the test statistic falls in the rejection region, the null hypothesis is rejected. There is sufficient evidence to indicate that the proportion of parents who describe their children as overweight is less than the actual proportion reported by the American Obesity Association.

c The p-value is calculated as

$$p\text{-value} = P(z < -4.53) < .0002 \quad \text{or} \quad p\text{-value} \approx 0.$$ Since the p-value is less than .05, the null hypothesis is rejected as in part **b**.

9.35   **a-b** Since the survival rate without screening is $p = 2/3$, the survival rate with an effective program may be greater than 2/3. Hence, the hypothesis to be tested is

$$H_0 : p = 2/3 \quad \text{versus} \quad H_a : p > 2/3$$

c   With $\hat{p} = \dfrac{x}{n} = \dfrac{164}{200} = .82$, the test statistic is

$$z = \frac{\hat{p} - p_0}{\sqrt{\dfrac{p_0 q_0}{n}}} = \frac{.82 - 2/3}{\sqrt{\dfrac{(2/3)(1/3)}{200}}} = 4.6$$

The rejection region is one-tailed, with $\alpha = .05$ or $z > 1.645$ and H_0 is rejected. The screening program seems to increase the survival rate.

d   For the one-tailed test,

$$p\text{-value} = P(z > 4.6) < 1 - .9998 = .0002$$

That is, H_0 can be rejected for any value of $\alpha \ge .0002$. The results are *highly significant.*

9.37   **a**   The hypothesis of interest is

$$H_0 : p = .5 \quad \text{versus} \quad H_a : p > .5$$

where p is the probability that the national brand is judged to be better than the store brand.

b   With $\hat{p} = \dfrac{x}{n} = \dfrac{8}{35} = .229$, the test statistic is

$$z = \frac{\hat{p} - p_0}{\sqrt{\dfrac{p_0 q_0}{n}}} = \frac{.229 - .5}{\sqrt{\dfrac{.5(.5)}{35}}} = -3.21$$

The rejection region is one-tailed with $\alpha = .01$, or $z > 2.33$ and H_0 is not rejected. The observed value of the test statistic actually ends up in the wrong tail of the distribution! There is insufficient evidence to indicate that the national brand is preferred to the store brand.

9.39   The hypothesis of interest is

$$H_0 : p = .80 \quad \text{versus} \quad H_a : p < .80$$

with $\hat{p} = \dfrac{x}{n} = \dfrac{37}{50} = .74$, the test statistic is

$$z = \frac{\hat{p} - p_0}{\sqrt{\dfrac{p_0 q_0}{n}}} = \frac{.74 - .80}{\sqrt{\dfrac{.80(.20)}{50}}} = -1.06$$

The rejection region with $\alpha = .05$ is $z < -1.645$ and the null hypothesis is not rejected. (Alternatively, we could calculate $p\text{-value} = P(z < -1.06) = .1446$. Since this p-value is greater than .05, the null hypothesis is not rejected.) There is insufficient evidence to refute the experimenter's claim.

9.41   The hypothesis of interest is

$$H_0 : p = .40 \quad \text{versus} \quad H_a : p \ne .40$$

with $\hat{p} = \dfrac{x}{n} = \dfrac{114}{300} = .38$, the test statistic is

$$z = \frac{\hat{p} - p_0}{\sqrt{\dfrac{p_0 q_0}{n}}} = \frac{.38 - .40}{\sqrt{\dfrac{.40(.60)}{300}}} = -.71$$

The rejection region with $\alpha = .05$ is $|z| > 1.96$ and the null hypothesis is not rejected. (Alternatively, we could calculate p-value $= 2P(z < -.71) = 2(.2389) = .4778$. Since this p-value is greater than .05, the null hypothesis is not rejected.) There is insufficient evidence to indicate that the proportion of households with at least one dog is different from that reported by the Humane Society.

9.43   **a-b** If p_1 cannot be larger than p_2, the only alternative to $H_0 : p_1 - p_2 = 0$ is that $p_1 < p_2$, and the one-tailed alternative is $H_a : p_1 - p_2 < 0$.

c The rejection region, with $\alpha = .05$, is $z < -1.645$ and the observed value of the test statistic is $z = -.84$. The null hypothesis is not rejected. There is no evidence to indicate that p_1 is smaller than p_2.

9.45   **a** The hypothesis of interest is:
$$H_0 : p_1 - p_2 = 0 \quad \text{versus} \quad H_a : p_1 - p_2 < 0$$
Calculate $\hat{p}_1 = .36$, $\hat{p}_2 = .60$ and $\hat{p} = \dfrac{n_1 \hat{p}_1 + n_2 \hat{p}_2}{n_1 + n_2} = \dfrac{18 + 30}{50 + 50} = .48$. The test statistic is then
$$z = \frac{\hat{p}_1 - \hat{p}_2}{\sqrt{\hat{p}\hat{q}\left(\dfrac{1}{n_1} + \dfrac{1}{n_2}\right)}} = \frac{.36 - .60}{\sqrt{.48(.52)(1/50 + 1/50)}} = -2.40$$

The rejection region, with $\alpha = .05$, is $z < -1.645$ and H_0 is rejected. There is evidence of a difference in the proportion of survivors for the two groups.

b From Section 8.7, the approximate 95% confidence interval is
$$(\hat{p}_1 - \hat{p}_2) \pm 1.96 \sqrt{\frac{\hat{p}_1 \hat{q}_1}{n_1} + \frac{\hat{p}_2 \hat{q}_2}{n_2}}$$
$$(.36 - .60) \pm 1.96 \sqrt{\frac{.36(.64)}{50} + \frac{.60(.40)}{50}}$$
$$-.24 \pm .19 \quad \text{or} \quad -.43 < (p_1 - p_2) < -.05$$

9.47   The hypothesis of interest is
$$H_0 : p_1 - p_2 = 0 \quad \text{versus} \quad H_a : p_1 - p_2 \neq 0$$
Calculate $\hat{p}_1 = \dfrac{12}{56} = .214$, $\hat{p}_2 = \dfrac{8}{32} = .25$, and $\hat{p} = \dfrac{x_1 + x_2}{n_1 + n_2} = \dfrac{12 + 8}{56 + 32} = .227$.
The test statistic is then
$$z = \frac{\hat{p}_1 - \hat{p}_2}{\sqrt{\hat{p}\hat{q}\left(\dfrac{1}{n_1} + \dfrac{1}{n_2}\right)}} = \frac{.214 - .25}{\sqrt{.227(.773)(1/56 + 1/32)}} = -.39$$

The rejection region, with $\alpha = .05$, is $|z| > 1.96$ and H_0 is not rejected. There is insufficient evidence to indicate a difference in the proportion of red M&Ms for the plain and peanut varieties. These results match the conclusions of Exercise 8.57.

9.49   Refer to Exercise 9.48. Calculate $\hat{p}_1 = \dfrac{40}{2266} = .018$ and $\hat{p}_2 = \dfrac{21}{2266} = .009$. The 99% lower one-sided confidence bound for $p_1 - p_2$ is

$$(\hat{p}_1 - \hat{p}_2) - 2.33\sqrt{\frac{\hat{p}_1\hat{q}_1}{n_1} + \frac{\hat{p}_2\hat{q}_2}{n_2}}$$

$$(.018-.009) - 2.33\sqrt{\frac{.018(.982)}{2266} + \frac{.009(.991)}{2266}}$$

$$.009 - .008 = .001 \quad \text{or} \quad (p_1 - p_2) > .001$$

The difference in risk between the two groups is at least 1 in 1000. This difference may not be of *practical significance* if the benefits of *Prempro* to the patient outweigh the risk.

9.51   The hypothesis of interest is
$$H_0 : p_1 - p_2 = 0 \quad \text{versus} \quad H_a : p_1 - p_2 > 0$$

Calculate $\hat{p}_1 = \frac{93}{121} = .769$, $\hat{p}_2 = \frac{119}{199} = .598$, and $\hat{p} = \frac{x_1 + x_2}{n_1 + n_2} = \frac{93 + 119}{121 + 199} = .6625$. The test statistic is then

$$z = \frac{\hat{p}_1 - \hat{p}_2}{\sqrt{\hat{p}\hat{q}\left(\frac{1}{n_1} + \frac{1}{n_2}\right)}} = \frac{.769 - .598}{\sqrt{.6625(.3375)(1/121 + 1/199)}} = 3.14$$

with p-value $= P(z > 3.14) = 1 - .9992 = .0008$. Since the p-value is less than .01, the results are reported as highly significant at the 1% level of significance. There is evidence to confirm the researcher's conclusion.

9.53   See Section 9.3 of the text.

9.55   The power of the test is $1 - \beta = P(\text{reject } H_0 \text{ when } H_0 \text{ is false})$. As μ gets farther from μ_0, the power of the test increases.

9.57   The objective of this experiment is to make a decision about the binomial parameter p, which is the probability that a customer prefers the first color. Hence, the null hypothesis will be that a customer has no preference for the first color, and the alternative will be that he does have a preference. If the null hypothesis is true, then
$$H_0 : p = P[\text{customer prefers the first color}] = 1/3$$
If the customer actually has a preference for the first color, then
$$H_a : p > 1/3$$

a   The test statistic is calculated with $\hat{p} = \frac{400}{1000} = .4$ as

$$z = \frac{\hat{p} - p_0}{\sqrt{\frac{p_0 q_0}{n}}} = \frac{.4 - (1/3)}{\sqrt{\frac{(1/3)(2/3)}{1000}}} = 4.47$$

and the p-value is
$$p\text{-value} = P(z > 4.47) < 1 - .9998 = .0002$$

since $P(z > 4.47)$ is surely less than $P(z > 3.49)$, the largest value in Table 3. (Alternatively, you could choose to report p-value ≈ 0.)

b   Since $\alpha = .05$ is larger than the p-value, which is less than .0002, H_0 can be rejected. We conclude that customers have a preference for the first color.

9.59   **a-b** Since it is necessary to prove that the average pH level is less than 7.5, the hypothesis to be tested is one-tailed:
$$H_0 : \mu = 7.5 \quad \text{versus} \quad H_a : \mu < 7.5$$

d The test statistic is

$$z = \frac{\bar{x} - \mu}{\sigma/\sqrt{n}} \approx \frac{\bar{x} - \mu}{s/\sqrt{n}} = \frac{-.2}{.2/\sqrt{30}} = -5.477$$

and the rejection region with $\alpha = .05$ is $z < -1.645$. The observed value, $z = -5.477$, falls in the rejection region and H_0 is rejected. We conclude that the average pH level is less than 7.5.

9.61 **a-b** Since there is no prior knowledge as to which mean should be larger, the hypothesis of interest is two-tailed

$$H_0 : \mu_1 - \mu_2 = 0 \quad \text{versus} \quad H_a : \mu_1 - \mu_2 \neq 0$$

c The test statistic is approximately

$$z \approx \frac{(\bar{x}_1 - \bar{x}_2) - 0}{\sqrt{\dfrac{s_1^2}{n_1} + \dfrac{s_2^2}{n_2}}} = \frac{2980 - 3205}{\sqrt{\dfrac{1140^2}{40} + \dfrac{963^2}{40}}} = -.954$$

The rejection region, with $\alpha = .05$, is two-tailed or $|z| > 1.96$. The null hypothesis is not rejected. There is insufficient evidence to indicate a difference in the two means.

9.63 Let p_1 be the proportion of defectives produced by machine A and p_2 be the proportion of defectives produced by machine B. The hypothesis to be tested is

$$H_0 : p_1 - p_2 = 0 \quad \text{versus} \quad H_a : p_1 - p_2 \neq 0$$

Calculate $\hat{p}_1 = \dfrac{16}{200} = .08$, $\hat{p}_2 = \dfrac{8}{200} = .04$, and $\hat{p} = \dfrac{x_1 + x_2}{n_1 + n_2} = \dfrac{16 + 8}{200 + 200} = .06$. The test statistic is then

$$z = \frac{\hat{p}_1 - \hat{p}_2}{\sqrt{\hat{p}\hat{q}\left(\dfrac{1}{n_1} + \dfrac{1}{n_2}\right)}} = \frac{.08 - .04}{\sqrt{.06(.94)(1/200 + 1/200)}} = 1.684$$

The rejection region, with $\alpha = .05$, is $|z| > 1.96$ and H_0 is not rejected. There is insufficient evidence to indicate that the machines are performing differently in terms of the percentage of defectives being produced.

9.65 **a-b** The hypothesis to be tested is

$$H_0 : p_1 - p_2 = 0 \quad \text{versus} \quad H_a : p_1 - p_2 > 0$$

Calculate $\hat{p}_1 = \dfrac{42}{66} = .636$, $\hat{p}_2 = \dfrac{48}{77} = .623$, and $\hat{p} = \dfrac{x_1 + x_2}{n_1 + n_2} = \dfrac{42 + 48}{66 + 77} = .629$. The test statistic is then

$$z = \frac{\hat{p}_1 - \hat{p}_2}{\sqrt{\hat{p}\hat{q}\left(\dfrac{1}{n_1} + \dfrac{1}{n_2}\right)}} = \frac{.636 - .623}{\sqrt{.629(.371)(1/66 + 1/77)}} = 0.16$$

and the p-value is $P(z \geq 0.16) = 1 - .5636 = .4364$.

c Since the observed p-value, .4364, is greater than $\alpha = .05$, H_0 cannot be rejected. There is insufficient evidence to support the researcher's belief.

9.67 The hypothesis to be tested is

$$H_0 : \mu_1 - \mu_2 = 0 \quad \text{versus} \quad H_a : \mu_1 - \mu_2 > 0$$

and the test statistic is

$$z \approx \frac{(\bar{x}_1 - \bar{x}_2) - 0}{\sqrt{\dfrac{s_1^2}{n_1} + \dfrac{s_2^2}{n_2}}} = \frac{10 - 8}{\sqrt{\dfrac{4.3}{40} + \dfrac{5.7}{40}}} = 4$$

The rejection region, with $\alpha = .05$, is one-tailed or $z > 1.645$ and the null hypothesis is rejected. There is sufficient evidence to indicate a difference in the two means. Hence, we conclude that diet I has a greater mean weight loss than diet II.

9.69 The hypothesis to be tested is

$$H_0 : \mu_1 - \mu_2 = 0 \quad \text{versus} \quad H_a : \mu_1 - \mu_2 \neq 0$$

and the test statistic is

$$z \approx \frac{(\overline{x}_1 - \overline{x}_2) - 0}{\sqrt{\dfrac{s_1^2}{n_1} + \dfrac{s_2^2}{n_2}}} = \frac{1925 - 1905}{\sqrt{\dfrac{40^2}{100} + \dfrac{30^2}{100}}} = 4$$

The rejection region, with $\alpha = .05$, is two-tailed or $|z| > 1.96$ and the null hypothesis is rejected. There is difference in mean breaking strengths for the two cables.

9.71 **a** The hypothesis to be tested is

$$H_0 : \mu_1 - \mu_2 = 0 \quad \text{versus} \quad H_a : \mu_1 - \mu_2 > 0$$

and the test statistic is

$$z \approx \frac{(\overline{x}_1 - \overline{x}_2) - 0}{\sqrt{\dfrac{s_1^2}{n_1} + \dfrac{s_2^2}{n_2}}} = \frac{240 - 227}{\sqrt{\dfrac{980}{200} + \dfrac{820}{200}}} = 4.33$$

The rejection region, with $\alpha = .05$, is one-tailed or $z > 1.645$ and the null hypothesis is rejected. There is a difference in mean yield for the two types of spray.

b An approximate 95% confidence interval for $\mu_1 - \mu_2$ is

$$(\overline{x}_1 - \overline{x}_2) \pm 1.96 \sqrt{\frac{s_1^2}{n_1} + \frac{s_2^2}{n_2}}$$

$$(240 - 227) \pm 1.96 \sqrt{\frac{980}{200} + \frac{820}{200}}$$

$$13 \pm 5.88 \quad \text{or} \quad 7.12 < (\mu_1 - \mu_2) < 18.88$$

9.73 **a** The hypothesis to be tested is

$$H_0 : \mu = 501 \quad \text{versus} \quad H_a : \mu \neq 501.$$

The test statistic is

$$z \approx \frac{\overline{x} - \mu}{s/\sqrt{n}} = \frac{499 - 501}{98/\sqrt{100}} = -.20$$

and the p-value is

$$p\text{-value} = P(z > .20) + P(z < -.20) = 2(.4207) = .8414$$

Since the p-value, .8414, is greater than $\alpha = .05$, and H_0 cannot be rejected and we cannot conclude that the average critical reading score for California students in 2010 is different from the national average.

b The hypothesis to be tested is

$$H_0 : \mu = 516 \quad \text{versus} \quad H_a : \mu \neq 516.$$

The test statistic is

$$z \approx \frac{\overline{x} - \mu}{s/\sqrt{n}} = \frac{514 - 516}{96/\sqrt{100}} = -.21$$

and the p-value is

$$p\text{-value} = P(z > .21) + P(z < -.21) = 2(.4168) = .8336$$

112

Since the p-value, .8336, is greater than $\alpha = .05$, and H_0 cannot be rejected and we cannot conclude that the average math score for California students in 2010 is different from the national average.

c Since the same students are used to measured critical reading and math scores, there would not be two independent samples, and the two sample z-test would not be appropriate.

9.75 The hypothesis to be tested is

$$H_0 : \mu = 5 \quad \text{versus} \quad H_a : \mu > 5$$

and the test statistic is

$$z = \frac{\bar{x} - \mu_0}{\sigma/\sqrt{n}} \approx \frac{\bar{x} - \mu_0}{s/\sqrt{n}} = \frac{7.2 - 5}{6.2/\sqrt{38}} = 2.19$$

The rejection region with $\alpha = .01$ is $z > 2.33$. Since the observed value, $z = 2.19$, does not fall in the rejection region and H_0 is not rejected. The data do not provide sufficient evidence to indicate that the mean ppm of PCBs in the population of game birds exceeds the FDA's recommended limit of 5 ppm.

9.77 The hypothesis of interest is

$$H_0 : p = .43 \quad \text{versus} \quad H_a : p \neq .43$$

with $\hat{p} = \dfrac{x}{n} = \dfrac{46}{100} = .46$, the test statistic is

$$z = \frac{\hat{p} - p_0}{\sqrt{\dfrac{p_0 q_0}{n}}} = \frac{.46 - .43}{\sqrt{\dfrac{.43(.57)}{100}}} = 0.61$$

The rejection region, with $\alpha = .05$, is two-tailed or $|z| > 1.96$ and the null hypothesis is not rejected. There is insufficient evidence to indicate that the percentage reported by *USA Today* is incorrect.

9.79 The hypothesis to be tested is

$$H_0 : p_1 - p_2 = 0 \quad \text{versus} \quad H_a : p_1 - p_2 \neq 0$$

Calculate $\hat{p}_1 = \dfrac{x_1}{602} = .54$, $\hat{p}_2 = \dfrac{x_2}{459} = .31$, and $\hat{p} = \dfrac{x_1 + x_2}{n_1 + n_2} = \dfrac{602(.54) + 459(.31)}{602 + 459} = .4405$. The test statistic is then

$$z = \frac{\hat{p}_1 - \hat{p}_2}{\sqrt{\hat{p}\hat{q}\left(\dfrac{1}{n_1} + \dfrac{1}{n_2}\right)}} = \frac{.54 - .31}{\sqrt{.4405(.5595)(1/602 + 1/459)}} = 7.48$$

The rejection region for $\alpha = .01$ is $|z| > 2.58$ and the null hypothesis is rejected. There is sufficient evidence to indicate that the proportion of students who are Advanced or Early-Advanced in English proficiency differs for these two districts.

9.81 **a** An approximate 99% confidence interval for $\mu_1 - \mu_2$ is

$$(\bar{x}_1 - \bar{x}_2) \pm 2.58 \sqrt{\frac{s_1^2}{n_1} + \frac{s_2^2}{n_2}}$$

$$(9017 - 5853) \pm 2.58 \sqrt{\frac{7162^2}{130} + \frac{1961^2}{80}}$$

$$3164 \pm 1716.51 \quad \text{or} \quad 1447.49 < (\mu_1 - \mu_2) < 4880.51$$

b Pure breaststroke swimmer swim between 1447 and 4881 more meters per week than to individual medley swimmers. This would be reasonable, since swimmers in the individual medley have three other strokes to practice—freestyle, backstroke and butterfly.

10: Inference from Small Samples

10.1 Refer to Table 4, Appendix I, indexing df along the left or right margin and t_α across the top.

   a $t_{.05} = 2.015$ with 5 df **b** $t_{.025} = 2.306$ with 8 df

   c $t_{.10} = 1.330$ with 18 df **d** $t_{.025} \approx 1.96$ with 30 df

10.3 **a** The p-value for a two-tailed test is defined as
$$p\text{-value} = P(|t| > 2.43) = 2P(t > 2.43)$$

so that
$$P(t > 2.43) = \frac{1}{2}\,p\text{-value}$$

Refer to Table 4, Appendix I, with $df = 12$. Remember that the value $P(t > t_a) = a$ is the tabled entry for a particular number of degrees of freedom. The exact probability, $P(t > 2.43)$ is unavailable; however, it is evident that $t = 2.43$ falls between $t_{.025} = 2.179$ and $t_{.01} = 2.681$. Therefore, the area to the right of $t = 2.43$ must be between .01 and .025. Since
$$.01 < \frac{1}{2}\,p\text{-value} < .025$$

the p-value can be approximated as
$$.02 < p\text{-value} < .05$$

   b For a right-tailed test, $p\text{-value} = P(t > 3.21)$ with $df = 16$. Since the value $t = 3.21$ is larger than $t_{.005} = 2.921$, the area to its right must be less than .005 and you can bound the p-value as
$$p\text{-value} < .005$$

   c For a two-tailed test, $p\text{-value} = P(|t| > 1.19) = 2P(t > 1.19)$, so that $P(t > 1.19) = \frac{1}{2}\,p\text{-value}$. From Table 4 with $df = 25$, $t = 1.19$ is smaller than $t_{.10} = 1.316$ so that
$$\frac{1}{2}\,p\text{-value} > .10 \quad \text{and} \quad p\text{-value} > .20$$

   d For a left-tailed test, $p\text{-value} = P(t < -8.77) = P(t > 8.77)$ with $df = 7$. Since the value $t = 8.77$ is larger than $t_{.005} = 3.499$, the area to its right must be less than .005 and you can bound the p-value as
$$p\text{-value} < .005$$

10.5 **a** Using the formulas given in Chapter 2, calculate $\sum x_i = 70.5$ and $\sum x_i^2 = 499.27$. Then
$$\bar{x} = \frac{\sum x_i}{n} = \frac{70.5}{10} = 7.05$$

$$s^2 = \frac{\sum x_i^2 - \frac{(\sum x_i)^2}{n}}{n-1} = \frac{499.27 - \frac{(70.5)^2}{10}}{9} = .249444 \quad \text{and} \quad s = .4994$$

   b Small sample confidence intervals are quite similar to their large sample counterparts; however, these intervals must be based on the *t-distribution*. With $df = n-1 = 9$, the appropriate value of t is $t_{.01} = 2.821$ (from Table 4) and the 99% upper one-sided confidence bound is
$$\bar{x} + t_{.01}\frac{s}{\sqrt{n}} \implies 7.05 + 2.821\sqrt{\frac{.249444}{10}} \implies 7.05 + .446$$

114

or $\mu < 7.496$. Intervals constructed using this procedure will enclose μ 99% of the time in repeated sampling. Hence, we are fairly certain that this particular interval encloses μ.

c The hypothesis to be tested is

$$H_0 : \mu = 7.5 \quad \text{versus} \quad H_a : \mu < 7.5$$

and the test statistic is

$$t = \frac{\bar{x} - \mu}{s/\sqrt{n}} = \frac{7.05 - 7.5}{\sqrt{\dfrac{.249444}{10}}} = -2.849$$

The rejection region with $\alpha = .01$ and $n - 1 = 9$ degrees of freedom is located in the lower tail of the t-distribution and is found from Table 4 as $t < -t_{.01} = -2.821$. Since the observed value of the test statistic falls in the rejection region, H_0 is rejected and we conclude that μ is less than 7.5.

d Notice that the 99% upper one-sided confidence bound for μ does not include the value $\mu = 7.5$. This would confirm the results of the hypothesis test in part **c**, in which we concluded that μ is less than 7.5.

10.7 Similar to Exercise 10.5. The hypothesis to be tested is

$$H_0 : \mu = 5 \quad \text{versus} \quad H_a : \mu < 5$$

Calculate $\bar{x} = \dfrac{\sum x_i}{n} = \dfrac{29.6}{6} = 4.933$

$$s^2 = \frac{\sum x_i^2 - \dfrac{\left(\sum x_i \right)^2}{n}}{n - 1} = \frac{146.12 - \dfrac{\left(29.6 \right)^2}{6}}{5} = .01867 \quad \text{and} \quad s = .1366$$

The test statistic is

$$t = \frac{\bar{x} - \mu}{s/\sqrt{n}} = \frac{4.933 - 5}{\dfrac{.1366}{\sqrt{6}}} = -1.195$$

The critical value of t with $\alpha = .05$ and $n - 1 = 5$ degrees of freedom is $t_{.05} = 2.015$ and the rejection region is $t < -2.015$. Since the observed value does not fall in the rejection region, H_0 is not rejected. There is no evidence to indicate that the dissolved oxygen content is less than 5 parts per million.

10.9 **a** Similar to previous exercises. The hypothesis to be tested is

$$H_0 : \mu = 100 \quad \text{versus} \quad H_a : \mu < 100$$

Calculate $\bar{x} = \dfrac{\sum x_i}{n} = \dfrac{1797.095}{20} = 89.85475$

$$s^2 = \frac{\sum x_i^2 - \dfrac{\left(\sum x_i \right)^2}{n}}{n - 1} = \frac{165,697.7081 - \dfrac{\left(1797.095 \right)^2}{20}}{19} = 222.1150605 \quad \text{and} \quad s = 14.9035$$

The test statistic is

$$t = \frac{\bar{x} - \mu}{s/\sqrt{n}} = \frac{89.85475 - 100}{\dfrac{14.9035}{\sqrt{20}}} = -3.044$$

The critical value of t with $\alpha = .01$ and $n - 1 = 19$ degrees of freedom is $t_{.01} = 2.539$ and the rejection region is $t < -2.539$. The null hypothesis is rejected and we conclude that μ is less than 100 DL.

b The 95% upper one-sided confidence bound, based on $n - 1 = 19$ degrees of freedom, is

$$\bar{x} + t_{.05} \frac{s}{\sqrt{n}} \Rightarrow 89.85475 + 2.539 \frac{14.90352511}{\sqrt{20}} \Rightarrow \mu < 98.316$$

This confirms the results of part **a** in which we concluded that the mean is less than 100 DL.

10.11 Calculate $\bar{x} = \dfrac{\sum x_i}{n} = \dfrac{37.82}{10} = 3.782$

$$s^2 = \frac{\sum x_i^2 - \dfrac{(\sum x_i)^2}{n}}{n-1} = \frac{143.3308 - \dfrac{(37.82)^2}{10}}{9} = .03284 \quad \text{and} \quad s = .1812$$

The 95% confidence interval based on $df = 9$ is

$$\bar{x} \pm t_{.025}\frac{s}{\sqrt{n}} \;\Rightarrow\; 3.782 \pm 2.262\frac{.1812}{\sqrt{10}} \;\Rightarrow\; 3.782 \pm .130$$

or $3.652 < \mu < 3.912$.

10.13 **a** The hypothesis to be tested is

$$H_0 : \mu = 25 \quad \text{versus} \quad H_a : \mu < 25$$

The test statistic is

$$t = \frac{\bar{x} - \mu_0}{s/\sqrt{n}} = \frac{20.3 - 25}{\dfrac{5}{\sqrt{21}}} = -4.31$$

The critical value of t with $\alpha = .05$ and $n-1 = 20$ degrees of freedom is $t_{.05} = 1.725$ and the rejection region is $t < -1.725$. Since the observed value does falls in the rejection region, H_0 is rejected, and we conclude that pre-treatment mean is less than 25.

b The 95% confidence interval based on $df = 20$ is

$$\bar{x} \pm t_{.025}\frac{s}{\sqrt{n}} \;\Rightarrow\; 26.6 \pm 2.086\frac{7.4}{\sqrt{21}} \;\Rightarrow\; 26.6 \pm 3.37$$

or $23.23 < \mu < 29.97$.

c The pre-treatment mean looks considerably smaller than the other two means.

10.15 **a** The t test of the hypothesis

$$H_0 : \mu = 1 \quad \text{versus} \quad H_a : \mu \neq 1$$

is not significant, since the p-value $= .113$ associated with the test statistic

$$t = \frac{\bar{x} - \mu}{s/\sqrt{n}} = \frac{1.0522 - 1}{.1657/\sqrt{27}} = 1.64$$

is greater than .10. There is insufficient evidence to indicate that the mean weight per package is different from one-pound.

b In fact, the 95% confidence limits for the average weight per package are

$$\bar{x} \pm t_{.025}\frac{s}{\sqrt{n}} \;\Rightarrow\; 1.0522 \pm 2.056\frac{.1657}{\sqrt{27}} \;\Rightarrow\; 1.0522 \pm .06556$$

or $.98664 < \mu < 1.11776$. These values agree (except in the last decimal place) with those given in the printout. Remember that you used the rounded values of $\bar{x}$ and s from the printout, causing a small rounding error in the results.

10.17 Refer to Exercise 10.16, where we found $\bar{x} = 246.96$ and $s = 46.8244$. If we use the large sample method of Chapter 8, the large sample confidence interval is

$$\bar{x} \pm z_{.025}\frac{s}{\sqrt{n}} \;\Rightarrow\; 246.96 \pm 1.96\frac{46.8244}{\sqrt{50}} \;\Rightarrow\; 246.96 \pm 12.98$$

or $233.98 < \mu < 259.94$. The intervals in Exercises 10.16 and 10.17 are fairly similar, which is why we choose to approximate the sampling distribution of $\dfrac{\bar{x} - \mu}{s/\sqrt{n}}$ with a z distribution when $n > 30$.

10.19 **a** $s^2 = \dfrac{(n_1-1)s_1^2 + (n_2-1)s_2^2}{n_1+n_2-2} = \dfrac{9(3.4)+3(4.9)}{10+4-2} = 3.775$

       b $s^2 = \dfrac{(n_1-1)s_1^2 + (n_2-1)s_2^2}{n_1+n_2-2} = \dfrac{11(18)+20(23)}{12+21-2} = 21.2258$

10.21 **a** The hypothesis to be tested is: $H_0 : \mu_1 - \mu_2 = 0$ versus $H_a : \mu_1 - \mu_2 \neq 0$

       b The rejection region is two-tailed, based on $df = n_1 + n_2 - 2 = 16 + 13 - 2 = 27$ degrees of freedom. With $\alpha = .01$, from Table 4, the rejection region is $|t| > t_{.005} = 2.771$.

       c The pooled estimator of σ^2 is calculated as

$$s^2 = \frac{(n_1-1)s_1^2 + (n_2-1)s_2^2}{n_1+n_2-2} = \frac{15(4.8)+12(5.9)}{16+13-2} = 5.2889$$

and the test statistic is

$$t = \frac{(\bar{x}_1 - \bar{x}_2) - 0}{\sqrt{s^2\left(\dfrac{1}{n_1} + \dfrac{1}{n_2}\right)}} = \frac{34.6 - 32.2}{\sqrt{5.2889\left(\dfrac{1}{16} + \dfrac{1}{13}\right)}} = 2.795$$

       d The p-value is

$$p\text{-value} = P(|t| > 2.795) = 2P(t > 2.795), \text{ so that } P(t > 2.795) = \frac{1}{2}\,p\text{-value}.$$

From Table 4 with $df = 27$, $t = 2.795$ is greater than the largest tabulated value ($t_{.005} = 2.771$). Therefore, the area to the right of $t = 2.795$ must be less than .005 so that

$$\frac{1}{2}\,p\text{-value} < .005 \quad \text{and} \quad p\text{-value} < .01$$

       e Comparing the observed $t = 2.795$ to the critical value $t_{.005} = 2.771$ or comparing the p-value ($< .01$) to $\alpha = .01$, H_0 is rejected and we conclude that $\mu_1 \neq \mu_2$.

10.23 **a** If you check the ratio of the two variances using the rule of thumb given in this section you will find:

$$\frac{\text{larger } s^2}{\text{smaller } s^2} = \frac{(4.67)^2}{(4.00)^2} = 1.36$$

which is less than three. Therefore, it is reasonable to assume that the two population variances are equal.

       b From the *Minitab* printout, the test statistic is $t = .06$ with p-value $= .95$.

       c The value of $s = 4.38$ is labeled "Pooled StDev" in the printout, so that $s^2 = (4.38)^2 = 19.1844$.

       d Since the p-value $= .95$ is greater than .10, the results are not significant. There is insufficient evidence to indicate a difference in the two population means.

       e A 95% confidence interval for $(\mu_1 - \mu_2)$ is given as

$$(\bar{x}_1 - \bar{x}_2) \pm t_{.025}\sqrt{s^2\left(\frac{1}{n_1} + \frac{1}{n_2}\right)}$$

$$(29 - 28.86) \pm 2.201\sqrt{19.1844\left(\frac{1}{6} + \frac{1}{7}\right)}$$

$$.14 \pm 5.363 \quad \text{or} \quad -5.223 < (\mu_1 - \mu_2) < 5.503$$

Since the value $\mu_1 - \mu_2 = 0$ falls in the confidence interval, it is possible that the two population means are the same. There insufficient evidence to indicate a difference in the two population means.

117

10.25 **a** If the antiplaque rinse is effective, the plaque buildup should be less for the group using the antiplaque rinse. Hence, the hypothesis to be tested is $H_0 : \mu_1 - \mu_2 = 0$ versus $H_a : \mu_1 - \mu_2 > 0$.

b The pooled estimator of σ^2 is calculated as

$$s^2 = \frac{(n_1 - 1)s_1^2 + (n_2 - 1)s_2^2}{n_1 + n_2 - 2} = \frac{6(.32)^2 + 6(.32)^2}{7 + 7 - 2} = .1024$$

and the test statistic is

$$t = \frac{(\bar{x}_1 - \bar{x}_2) - 0}{\sqrt{s^2\left(\dfrac{1}{n_1} + \dfrac{1}{n_2}\right)}} = \frac{1.26 - .78}{\sqrt{.1024\left(\dfrac{1}{7} + \dfrac{1}{7}\right)}} = 2.806$$

The rejection region is one-tailed, based on $n_1 + n_2 - 2 = 12$ degrees of freedom. With $\alpha = .05$, from Table 4, the rejection region is $t > t_{.05} = 1.782$ and H_0 is rejected. There is evidence to indicate that the rinse is effective.

c The p-value is

$$p\text{-value} = P(t > 2.806)$$

From Table 4 with $df = 12$, $t = 2.806$ is between two tabled entries $t_{.005} = 3.055$ and $t_{.01} = 2.681$, we can conclude that

$$.005 < p\text{-value} < .01$$

10.27 **a** The hypothesis to be tested is

$$H_0 : \mu_1 - \mu_2 = 0 \quad \text{versus} \quad H_a : \mu_1 - \mu_2 \ne 0$$

where μ_1 is the average compartment pressure for runners, and μ_2 is the average compartment pressure for cyclists. The pooled estimator of σ^2 is calculated as

$$s^2 = \frac{(n_1 - 1)s_1^2 + (n_2 - 1)s_2^2}{n_1 + n_2 - 2} = \frac{9(3.92)^2 + 9(3.98)^2}{18} = 15.6034$$

and the test statistic is

$$t = \frac{(\bar{x}_1 - \bar{x}_2) - 0}{\sqrt{s^2\left(\dfrac{1}{n_1} + \dfrac{1}{n_2}\right)}} = \frac{14.5 - 11.1}{\sqrt{15.6034\left(\dfrac{1}{10} + \dfrac{1}{10}\right)}} = 1.92$$

The rejection region is two-tailed, based on $df = 18$ degrees of freedom. With $\alpha = .05$, from Table 4, the rejection region is $|t| > t_{.025} = 2.101$. We do not reject H_0; there is insufficient evidence to indicate a difference in the means.

b Calculate

$$s^2 = \frac{(n_1 - 1)s_1^2 + (n_2 - 1)s_2^2}{n_1 + n_2 - 2} = \frac{9(3.49)^2 + 9(4.95)^2}{18} = 18.3413$$

A 95% confidence interval for $(\mu_1 - \mu_2)$ is given as

$$(\bar{x}_1 - \bar{x}_2) \pm t_{.025}\sqrt{s^2\left(\frac{1}{n_1} + \frac{1}{n_2}\right)}$$

$$(12.2 - 11.5) \pm 2.101\sqrt{18.3413\left(\frac{1}{10} + \frac{1}{10}\right)}$$

$$.7 \pm 4.02 \quad \text{or} \quad -3.32 < (\mu_1 - \mu_2) < 4.72$$

c Check the ratio of the two variances using the rule of thumb given in this section:

$$\frac{\text{larger } s^2}{\text{smaller } s^2} = \frac{(16.9)^2}{(4.47)^2} = 14.29$$

118

which is greater than three. Therefore, it is not reasonable to assume that the two population variances are equal. You should use the unpooled t test with Satterthwaite's approximation to the degrees of freedom.

10.29 **a** Use your scientific calculator or the computing formulas to find:

$$\bar{x}_1 = .0125 \qquad s_1^2 = .000002278 \qquad s_1 = .001509$$
$$\bar{x}_2 = .0138 \qquad s_2^2 = .000003733 \qquad s_2 = .001932$$

Since the ratio of the variances is less than 3, you can use the pooled t test, calculating

$$s^2 = \frac{(n_1-1)s_1^2 + (n_2-1)s_2^2}{n_1+n_2-2} = \frac{9(.000002278)+9(.000003733)}{18} = .000003006$$

and the test statistic is

$$t = \frac{(\bar{x}_1 - \bar{x}_2)-0}{\sqrt{s^2\left(\frac{1}{n_1}+\frac{1}{n_2}\right)}} = \frac{.0125-.0138}{\sqrt{s^2\left(\frac{1}{10}+\frac{1}{10}\right)}} = -1.68$$

For a two-tailed test with $df = 18$, the p-value can be bounded using Table 4 so that

$$.05 < \frac{1}{2}\,p\text{-value} < .10 \quad \text{or} \quad .10 < p\text{-value} < .20$$

Since the p-value is greater than .10, $H_0: \mu_1 - \mu_2 = 0$ is not rejected. There is insufficient evidence to indicate that there is a difference in the mean titanium contents for the two methods.

b A 95% confidence interval for $(\mu_1 - \mu_2)$ is given as

$$(\bar{x}_1 - \bar{x}_2) \pm t_{.025}\sqrt{s^2\left(\frac{1}{n_1}+\frac{1}{n_2}\right)}$$

$$(.0125 - .0138) \pm 2.101\sqrt{s^2\left(\frac{1}{10}+\frac{1}{10}\right)}$$

$$-.0013 \pm .0016 \quad \text{or} \quad -.0029 < (\mu_1 - \mu_2) < .0003$$

Since $\mu_1 - \mu_2 = 0$ falls in the confidence interval, the conclusion of part **a** is confirmed. *This particular data set is very susceptible to rounding error. You need to carry as much accuracy as possible to obtain accurate results.*

10.31 **a** The hypothesis of interest is

$$H_0: \mu_1 - \mu_2 = 0 \quad \text{versus} \quad H_a: \mu_1 - \mu_2 > 0$$

When the actual data are given and s^2 must be calculated, the calculation is done by using your scientific calculator to first obtain s_1 and s_2 and substituting into the formula for s^2. Notice that

$$(n-1)s^2 = \sum(x_i - \bar{x})^2 = \sum x_i^2 - \frac{(\sum x_i)^2}{n}$$

Hence, for the pooled estimator, we can eliminate some rounding error by calculating

$$s^2 = \frac{\sum x_{1i}^2 - \frac{(\sum x_{1i})^2}{n_1} + \sum x_{2i}^2 - \frac{(\sum x_{2i})^2}{n_2}}{n_1 + n_2 - 2}$$

and the preliminary calculations are as follows:

Sample 1 (Above)	Sample 2 (Below)
$\sum x_{1i} = 25$	$\sum x_{2i} = 24.3$
$\sum x_{1i}^2 = 125.1$	$\sum x_{2i}^2 = 118.15$
$n_1 = 5$	$n_2 = 5$

Then

$$s^2 = \frac{(n_1 - 1)s_1^2 + (n_2 - 1)s_2^2}{n_1 + n_2 - 2}$$

$$= \frac{125.1 - \frac{(25)^2}{5} + 118.15 - \frac{(24.3)^2}{5}}{5 + 5 - 2} = \frac{.1 + .052}{8} = .019$$

Also, $\quad \bar{x}_1 = \frac{25}{5} = 5 \quad$ and $\quad \bar{x}_2 = \frac{24.3}{5} = 4.86$

The test statistic is

$$t = \frac{(\bar{x}_1 - \bar{x}_2) - 0}{\sqrt{s^2\left(\frac{1}{n_1} + \frac{1}{n_2}\right)}} = \frac{5 - 4.86}{\sqrt{.019\left(\frac{1}{5} + \frac{1}{5}\right)}} = 1.606$$

For a one-tailed test with $df = 8$ and $\alpha = .05$ the rejection region is $t > t_{.05} = 1.86$, and H_0 is not rejected. There is insufficient evidence to indicate that the mean content of oxygen below town is less than the mean content above.

b   A 95% confidence interval for $(\mu_1 - \mu_2)$ is given as

$$(\bar{x}_1 - \bar{x}_2) \pm t_{.025}\sqrt{s^2\left(\frac{1}{n_1} + \frac{1}{n_2}\right)}$$

$$(5 - 4.86) \pm 2.306\sqrt{.019\left(\frac{1}{5} + \frac{1}{5}\right)}$$

$$.14 \pm .201 \quad \text{or} \quad -.061 < (\mu_1 - \mu_2) < .341$$

10.33   Refer to Exercise 10.32 for the preliminary calculations. A 95% lower one-sided confidence bound for $(\mu_1 - \mu_2)$ is then given as

$$(\bar{x}_1 - \bar{x}_2) - t_{.05}\sqrt{s^2\left(\frac{1}{n_1} + \frac{1}{n_2}\right)}$$

$$(59.646 - 59.627) - 1.734\sqrt{.03124722\left(\frac{1}{10} + \frac{1}{10}\right)}$$

$$.019 - .137 \quad \text{or} \quad (\mu_1 - \mu_2) > -.118$$

Since the value $\mu_1 - \mu_2 = 0$ is in the interval, it is possible that the two means might be equal. We do not have enough evidence to indicate that there is a difference in the means.

10.35   The hypothesis of interest is

$$H_0 : \mu_1 - \mu_2 = 0 \quad \text{versus} \quad H_a : \mu_1 - \mu_2 \neq 0$$

and the preliminary calculations are as follows:

Island Thorns	Ashley Rails
$\sum x_{1i} = 90.9$	$\sum x_{2i} = 86.6$
$\sum x_{1i}^2 = 1665.17$	$\sum x_{2i}^2 = 1510.92$
$\bar{x}_1 = \frac{90.9}{5} = 18.18$	$\bar{x}_2 = \frac{86.6}{5} = 17.32$
$s_1 = 1.775387$	$s_2 = 1.658915$
$n_1 = 5$	$n_2 = 5$

120

Then

$$s^2 = \frac{(n_1 - 1)s_1^2 + (n_2 - 1)s_2^2}{n_1 + n_2 - 2}$$

$$= \frac{4(1.775387^2) + 4(1.658915^2)}{5 + 5 - 2} = 2.951999$$

The test statistic is

$$t = \frac{(\bar{x}_1 - \bar{x}_2) - 0}{\sqrt{s^2\left(\frac{1}{n_1} + \frac{1}{n_2}\right)}} = \frac{18.18 - 17.32}{\sqrt{2.951999\left(\frac{1}{5} + \frac{1}{5}\right)}} = .79$$

The rejection region with $\alpha = .05$ and $df = 5 + 5 - 2 = 8$ is $|t| > 2.306$ and the null hypothesis is not rejected. There is insufficient evidence to indicate a difference in the average percentage of aluminum oxide at the two sites.

10.37 **a** The hypothesis of interest is

$$H_0 : \mu_1 - \mu_2 = 0 \quad \text{or} \quad H_0 : \mu_d = 0$$
$$H_a : \mu_1 - \mu_2 > 0 \quad \text{or} \quad H_a : \mu_d > 0$$

b The test statistic is

$$t = \frac{\bar{d} - \mu_d}{s_d / \sqrt{n}} = \frac{5.7 - 0}{\sqrt{\frac{256}{18}}} = 1.511$$

with $n - 1 = 17$ degrees of freedom. The rejection region is with $\alpha = .05$ is $t > t_{.05} = 1.740$, and H_0 is not rejected. We cannot conclude that $\mu_d > 0$.

10.39 **a** A paired-difference test is used, since the two samples are not independent (for any given city, Geico and 21st Century premiums will be related).
b The hypothesis of interest is

$$H_0 : \mu_1 - \mu_2 = 0 \quad \text{or} \quad H_0 : \mu_d = 0$$
$$H_a : \mu_1 - \mu_2 \neq 0 \quad \text{or} \quad H_a : \mu_d \neq 0$$

where μ_1 is the average for Geico insurance and μ_2 is the average cost for 21st Century insurance. The table of differences, along with the calculation of $\bar{d}$ and s_d, is presented below.

City	1	2	3	4	Totals
d_i	428	−51	−23	−257	97
d_i^2	183,184	2601	529	66,049	252,363

$$\bar{d} = \frac{\sum d_i}{n} = \frac{97}{4} = 24.25 \qquad \text{and}$$

$$s_d = \sqrt{\frac{\sum d_i^2 - \frac{(\sum d_i)^2}{n}}{n - 1}} = \sqrt{\frac{252,363 - \frac{(97)^2}{4}}{3}} = \sqrt{83,336.916667} = 288.681341$$

The test statistic is

$$t = \frac{\bar{d} - \mu_d}{s_d / \sqrt{n}} = \frac{24.25 - 0}{\frac{288.681341}{\sqrt{4}}} = .168$$

with $n - 1 = 3$ degrees of freedom. The rejection region with $\alpha = .01$ is $|t| > t_{.005} = 5.841$, and H_0 is not rejected. There is insufficient evidence to indicate a difference in the average premiums for Geico and 21st Century.

c   p-value $= P\left(|t| > .168\right) = 2P\left(t > .268\right)$. Since $t = .168$ is smaller than $t_{.10} = 1.638$,

$$\text{p-value} > 2(.10) \Rightarrow \text{p-value} > .20 .$$

d   A 99% confidence interval for $\mu_1 - \mu_2 = \mu_d$ is

$$\bar{d} \pm t_{.005}\frac{s_d}{\sqrt{n}} \quad \Rightarrow \quad 24.25 \pm 5.841\frac{288.681341}{\sqrt{4}} \quad \Rightarrow \quad 24.25 \pm 843.094$$

or $-818.844 < \left(\mu_1 - \mu_2\right) < 867.344$.

e   The four cities in the study were not necessarily a random sample of cities from throughout the United States. Therefore, you cannot make valid comparisons between Geico and 21st Century for the United States in general.

10.41   **a-b** The table of differences, along with the calculation of $\bar{d}$ and s_d^2, is presented below.

Week	1	2	3	4	Totals
d_i	−1.77	−15.03	−23.22	−27.05	−67.07

$$\bar{d} = \frac{\sum d_i}{n} = \frac{-67.07}{4} = -16.7675$$

$$s_d^2 = \frac{\sum d_i^2 - \frac{\left(\sum d_i\right)^2}{n}}{n-1} = \frac{1499.9047 - \frac{(-67.07)^2}{4}}{3} = 125.102825 \quad \text{and} \quad s_d = 11.1849$$

The hypothesis of interest is

$$\text{H}_0 : \mu_1 - \mu_2 = 0 \quad \text{or} \quad \text{H}_0 : \mu_d = 0$$
$$\text{H}_a : \mu_1 - \mu_2 \neq 0 \quad \text{or} \quad \text{H}_a : \mu_d \neq 0$$

and the test statistic is

$$t = \frac{\bar{d} - \mu_d}{s_d/\sqrt{n}} = \frac{-16.7675 - 0}{\frac{11.1849}{\sqrt{4}}} = -3.00$$

Since $t = -3.00$ with $df = n - 1 = 3$ falls between the two tabled values, $t_{.025}$ and $t_{.05}$,

$$.025 < \frac{1}{2}(\text{p-value}) < .05$$
$$.05 < \text{p-value} < .10$$

for this two tailed test and H$_0$ is not rejected. We cannot conclude that the means are different.

c   The 99% confidence interval for $\mu_1 - \mu_2 = \mu_d$ is

$$\bar{d} \pm t_{.005}\frac{s_d}{\sqrt{n}} \quad \Rightarrow \quad -16.7675 \pm 5.841\frac{11.1849}{\sqrt{4}} \quad \Rightarrow \quad -16.7675 \pm 32.666$$

or $-49.433 < \left(\mu_1 - \mu_2\right) < 15.899$.

10.43   If two measurements are taken on the same person, the measurements are not independent, and a paired analysis should be used.

10.45   A paired-difference test is used, since the two samples are not random and independent (within any sample, the dye 1 and dye 2 measurements are related). The hypothesis of interest is

$$\text{H}_0 : \mu_1 - \mu_2 = 0 \quad \text{H}_a : \mu_1 - \mu_2 \neq 0$$

The table of differences, along with the calculation of $\bar{d}$ and s_d^2, is presented below.

Sample	1	2	3	4	5	6	7	8	9	Total
d_i	2	1	−1	2	3	−1	0	2	2	10

$$\bar{d} = \frac{\sum d_i}{n} = \frac{10}{9} = 1.11$$

$$s_d^2 = \frac{\sum d_i^2 - \frac{(\sum d_i)^2}{n}}{n-1} = \frac{28 - \frac{(10)^2}{9}}{8} = 2.1111 \quad \text{and} \quad s_d = 1.452966$$

and the test statistic is

$$t = \frac{\bar{d} - \mu_d}{s_d / \sqrt{n}} = \frac{1.11 - 0}{\frac{1.452966}{\sqrt{9}}} = 2.29$$

A rejection region with $\alpha = .05$ and $df = n-1 = 8$ is $|t| > t_{.025} = 2.306$, and H$_0$ is not rejected at the 5% level of significance. We cannot conclude that there is a difference in the mean brightness scores.

10.47 The hypothesis of interest is

$$H_0 : \mu_1 - \mu_2 = 0 \quad \text{or} \quad H_0 : \mu_d = 0$$
$$H_a : \mu_1 - \mu_2 \neq 0 \quad \text{or} \quad H_a : \mu_d \neq 0$$

The table of differences is presented below. Use your scientific calculator to find $\bar{d}$ and s_d,

d_i	15	15	15	9	16	8	12	8	10	12
	9	4	10	4	17	13	4	7	7	10

Calculate $\bar{d} = 10.25$, $s_d = 4.051$ and the test statistic is

$$t = \frac{\bar{d} - \mu_d}{s_d / \sqrt{n}} = \frac{10.25 - 0}{\frac{4.051}{\sqrt{20}}} = 11.32$$

Since $t = 11.32$ with $df = n-1 = 19$ is greater than the largest tabled value $t_{.005}$,

$$p\text{-value} < .005$$

for this one-tailed test and H$_0$ is rejected. We can conclude that the average recall score is higher when imagery is used. The results are highly significant $(P < .01)$.

10.49 It is necessary to test

$$H_0 : \sigma^2 = 15 \quad \text{versus} \quad H_a : \sigma^2 > 15$$

This will be done using s^2, the sample variance, which is a good estimator for σ^2. Refer to Section 10.6 of the text and notice that the quantity

$$\chi^2 = \frac{(n-1)s^2}{\sigma^2}$$

possesses a chi-square distribution in repeated sampling. This distribution is shown below.

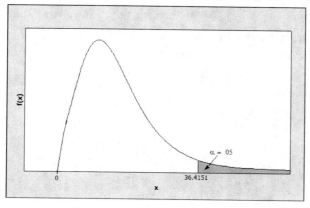

Notice that the distribution is nonsymmetrical and that the random variable

$$\frac{(n-1)s^2}{\sigma^2}$$

takes on values commencing at zero (since s^2, σ^2 and $(n-1)$ are never negative). The test statistic is

$$\chi^2 = \frac{(n-1)s^2}{\sigma_0^2} = \frac{24(21.4)}{15} = 34.24$$

A one-tailed test of hypothesis is required. Hence, a critical value of χ^2 (denoted by χ_C^2) must be found such that

$$P\left(\chi^2 > \chi_C^2\right) = .05$$

Indexing $\chi_{.05}^2$ with $n-1 = 24$ degrees of freedom (see Table 5), the critical value is found to be $\chi_{.05}^2 = 36.4151$ (see the figure on the previous page). The value of the test statistic does not fall in the rejection region. Hence, H_0 is not rejected. We cannot conclude that the variance exceeds 15.

10.51 **a** Calculate $\Sigma x_i = 17.7$, $\Sigma x_i^2 = 48.95$ and $n = 7$. Then

$$s^2 = \frac{\Sigma x_i^2 - \frac{(\Sigma x_i)^2}{n}}{n-1} = \frac{48.95 - \frac{(17.7)^2}{7}}{6} = .6990476$$

b Indexing $\chi_{.025}^2$ and $\chi_{.975}^2$ with $n-1 = 6$ degrees of freedom in Table 5 yields

$$\chi_{.025}^2 = 14.4494 \quad \text{and} \quad \chi_{.975}^2 = 1.237347$$

and the 95% confidence interval is

$$\frac{6(.6990476)}{14.4494} < \sigma^2 < \frac{6(.6990476)}{1.237347} \quad \text{or} \quad .291 < \sigma^2 < 3.390$$

c It is necessary to test

$$H_0 : \sigma^2 = .8 \quad \text{versus} \quad H_a : \sigma^2 \neq .8$$

and the test statistic is

$$\chi^2 = \frac{(n-1)s^2}{\sigma_0^2} = \frac{6(.6990476)}{.8} = 5.24$$

The two-tailed rejection region with $\alpha = .05$ and $n-1 = 6$ degrees of freedom is

$$\chi^2 > \chi_{.025}^2 = 14.4494 \quad \text{or} \quad \chi^2 < \chi_{.975}^2 = 1.237347$$

and H_0 is not rejected. There is insufficient evidence to indicate that σ^2 is different from .8.

d The p-value is found by approximating $P\left(\chi^2 > 5.24\right)$ and then doubling that value to account for an equally small value of s^2 which might have produced a value of the test statistic in the lower tail of the chi-square distribution. The observed value, $\chi^2 = 5.24$, is smaller than $\chi_{.10}^2 = 10.6646$ in Table 5. Hence,

$$p\text{-value} > 2(.10) = .20$$

10.53 **a** The hypothesis to be tested is

$$H_0 : \mu = 5 \quad H_a : \mu \neq 5$$

Calculate $\bar{x} = \frac{\Sigma x_i}{n} = \frac{19.96}{4} = 4.99$, $s^2 = \frac{\Sigma x_i^2 - \frac{(\Sigma x_i)^2}{n}}{n-1} = \frac{99.6226 - \frac{(19.96)^2}{4}}{3} = .0074$

and the test statistic is

$$t = \frac{\bar{x} - \mu_0}{s/\sqrt{n}} = \frac{4.99 - 5}{\sqrt{\frac{.0074}{4}}} = -.232$$

The rejection region with $\alpha = .05$ and $n - 1 = 3$ degrees of freedom is found from Table 4 as $|t| > t_{.025} = 3.182$. Since the observed value of the test statistic does not fall in the rejection region, H_0 is not rejected. There is insufficient evidence to show that the mean differs from 5 mg/cc.

b The manufacturer claims that the range of the potency measurements will equal .2. Since this range is given to equal 6σ, we know that $\sigma \approx .0333$. Then

$$H_0 : \sigma^2 = (.0333)^2 = .0011 \qquad H_a : \sigma^2 > .0011$$

The test statistic is

$$\chi^2 = \frac{(n-1)s^2}{\sigma_0^2} = \frac{3(.0074)}{.0011} = 20.18$$

and the one-tailed rejection region with $\alpha = .05$ and $n - 1 = 3$ degrees of freedom is

$$\chi^2 > \chi_{.05}^2 = 7.81$$

H_0 is rejected; there is sufficient evidence to indicate that the range of the potency will exceed the manufacturer's claim.

10.55 **a** The force transmitted to a wearer, x, is known to be normally distributed with $\mu = 800$ and $\sigma = 40$. Hence,

$$P(x > 1000) = P\left(z > \frac{1000 - 8000}{40}\right) = P(z > 5) \approx 0$$

It is highly improbable that any particular helmet will transmit a force in excess of 1000 pounds.

b Since $n = 40$, a large sample test will be used to test

$$H_0 : \mu = 800 \qquad H_a : \mu > 800$$

The test statistic is

$$t = \frac{\bar{x} - \mu_0}{s/\sqrt{n}} = \frac{825 - 800}{\sqrt{\frac{2350}{40}}} = 3.262$$

and the rejection region with $\alpha = .05$ is $z > 1.645$. H_0 is rejected and we conclude that $\mu > 800$.

10.57 The hypothesis of interest is $\qquad H_0 : \sigma = 150 \qquad H_a : \sigma < 150$
Calculate

$$(n-1)s^2 = \sum x_i^2 - \frac{(\sum x_i)^2}{n} = 92,305,600 - \frac{(42,812)^2}{20} = 662,232.8$$

and the test statistic is $\chi^2 = \frac{(n-1)s^2}{\sigma_0^2} = \frac{662,232.8}{150^2} = 29.433$. The one-tailed rejection region with

$\alpha = .01$ and $n - 1 = 19$ degrees of freedom is $\chi^2 < \chi_{.99}^2 = 7.63273$, and H_0 is not rejected. There is insufficient evidence to indicate that he is meeting his goal.

10.59 Refer to Exercise 10.58. When the assumptions for the F distribution are met, then s_1^2/s_2^2 possesses an F distribution with $df_1 = n_1 - 1$ and $df_2 = n_2 - 1$ degrees of freedom. Note that df_1 and df_2 are the degrees of freedom associated with s_1^2 and s_2^2, respectively. The F distribution is non-symmetrical with the degree of skewness dependent on the above-mentioned degrees of freedom. Table 6 presents the critical values of F (depending on the degrees of freedom) such that $P(F > F_a) = a$ for $a = .10, .05, .025, .01$ and $.005$, respectively. From Table 6, $F_{df_1, df_2} = 2.62$ and $F_{df_2, df_1} \approx 2.76$. The 95% confidence interval for σ_1^2/σ_2^2 is

$$\frac{s_1^2}{s_2^2}\frac{1}{F_{df_1,df_2}} < \frac{\sigma_1^2}{\sigma_2^2} < \frac{s_1^2}{s_2^2}F_{df_2,df_1}$$

$$\frac{55.7}{31.4}\left(\frac{1}{2.62}\right) < \frac{\sigma_1^2}{\sigma_2^2} < \frac{55.7}{31.4}(2.76) \quad \text{or} \quad .667 < \frac{\sigma_1^2}{\sigma_2^2} < 4.896$$

10.61 The hypothesis of interest is $\quad H_0 : \sigma_1^2 = \sigma_2^2 \quad$ versus $\quad H_a : \sigma_1^2 \neq \sigma_2^2$ and the test statistic is

$$F = \frac{s_1^2}{s_2^2} = \frac{114^2}{103^2} = 1.22 .$$

The critical values of F for various values of α are given below using $df_1 = 15$ and $df_2 = 14$.

α	.10	.05	.025	.01	.005
F_α	2.01	2.46	2.95	3.66	4.25

Hence,

$$p\text{-value} = 2P(F > 1.22) > 2(.10) = .20$$

Since the p-value is so large, H_0 is not rejected. There is no evidence to indicate that the variances are different.

10.63 **a** The hypothesis of interest is: $\quad H_0 : \sigma_1^2 = \sigma_2^2 \quad$ versus $\quad H_a : \sigma_1^2 \neq \sigma_2^2$

Calculate $s_1^2 = \dfrac{7106 - \dfrac{(312)^2}{15}}{14} = 44.02857$ and $s_2^2 = \dfrac{4982 - \dfrac{(240)^2}{12}}{11} = 16.545455$.

Then the test statistic is

$$F = \frac{s_1^2}{s_2^2} = \frac{44.02857}{16.545455} = 2.66 .$$

The rejection region is two-tailed, with $\alpha = .01$ and $df_1 = 14$ and $df_2 = 11$ degrees of freedom. The upper tail of this rejection region is approximated using $df_1 \approx 12$ and $df_2 = 11$ in Table 6 as $F > F_{.005} \approx 5.24$. Since the observed value of the test statistic does not fall in the rejection region, H_0 is not rejected. There is insufficient evidence to indicate a difference in the variability for the two quarterbacks.
b Since the F-test did not show any reason to doubt the equality of variance assumption, it is appropriate to use the two sample t-test which assumes equal variances.

10.65 For each of the three tests, the hypothesis of interest is

$$H_0 : \sigma_1^2 = \sigma_2^2 \quad \text{versus} \quad H_a : \sigma_1^2 \neq \sigma_2^2$$

and the test statistics are

$$F = \frac{s_1^2}{s_2^2} = \frac{3.98^2}{3.92^2} = 1.03 \qquad F = \frac{s_1^2}{s_2^2} = \frac{4.95^2}{3.49^2} = 2.01 \quad \text{and} \quad F = \frac{s_1^2}{s_2^2} = \frac{16.9^2}{4.47^2} = 14.29$$

The critical values of F for various values of α are given below using $df_1 = 9$ and $df_2 = 9$.

α	.10	.05	.025	.01	.005
F_α	2.44	3.18	4.03	5.35	6.54

Hence, for the first two tests,

$$p\text{-value} > 2(.10) = .20$$

while for the last test,

$$p\text{-value} < 2(.005) = .01$$

There is no evidence to indicate that the variances are different for the first two tests, but H_0 is rejected for the third variable. The two-sample t-test with a pooled estimate of σ^2 cannot be used for the third variable.

10.67 A Student's t test can be employed to test the hypothesis about a single population mean when the sample has been randomly selected from a normal population. It will work quite satisfactorily for populations which possess mound-shaped frequency distributions resembling the normal distribution.

10.69 Paired observations are used to estimate the difference between two population means in preference to an estimation based on independent random samples selected from the two populations because of the increased information caused by blocking the observations. We expect blocking to create a large reduction in the standard deviation, if differences do exist among the blocks.

Paired observations are not always preferable. The degrees of freedom that are available for estimating σ^2 are less for paired than for unpaired observations. If there were no difference between the blocks, the paired experiment would then be less beneficial.

10.71 **a** Use Table 4 with $df = 5$. Since the value $t = 1.2$ is smaller than $t_{.10} = 1.476$, the area to its right must be greater than .10 and you can bound the p-value as
$$p\text{-value} > .10$$

b For a two-tailed test with $df = 10$, $p\text{-value} = P(t > 2) + P(t < -2) = 2P(t > 2)$, so that

$$P(t > 2) = \frac{1}{2} p\text{-value}.$$ From Table 4 with $df = 10$, $t = 2$ falls between $t_{.025} = 2.228$ and $t_{.05} = 1.812$ so that

$$.025 < \frac{1}{2} p\text{-value} < .05 \quad \text{and} \quad .05 < p\text{-value} < .10$$

c Use Table 4 with $df = 8$, and remember that $P(t < -3.3) = P(t > 3.3)$. Since the value $t = 3.3$ falls between $t_{.01} = 2.896$ and $t_{.005} = 3.355$, you can bound the p-value as
$$.005 < p\text{-value} < .01$$

d Use Table 4 with $df = 12$. Since the value $t = 0.6$ is smaller than $t_{.10} = 1.356$, the area to its right must be greater than .10 and you can bound the p-value as
$$p\text{-value} > .10$$

10.73 **a** The hypothesis to be tested is
$$H_0 : \mu = .05 \qquad H_a : \mu > .05$$

and the test statistic is

$$t = \frac{\bar{x} - \mu_0}{s/\sqrt{n}} = \frac{.058 - .05}{\frac{.012}{\sqrt{10}}} = 2.108.$$

Critical value approach: The rejection region with $\alpha = .05$ and $df = n - 1 = 9$ degrees of freedom is located in the upper tail of the t-distribution and is found from Table 4 as $t > t_{.05} = 1.833$. Since the observed value falls in the rejection region, H_0 is rejected and we conclude that μ is greater than .05.

p-value approach: The $p\text{-value} = P(t > 2.108)$. Since the value $t = 2.108$ falls between $t_{.025}$ and $t_{.05}$, the p-value can be bounded as
$$.025 < p\text{-value} < .05$$
In any event, the p-value is less than .05 and H_0 can be rejected at the 5% level of significance.

10.75 Using the formulas given in Chapter 2 or your scientific calculator, calculate
$$\bar{x} = \frac{\sum x_i}{n} = \frac{233.29}{3} = 77.763$$

$$s^2 = \frac{\sum x_i^2 - \frac{(\sum x_i)^2}{n}}{n-1} = \frac{18169.666 - \frac{(233.29)^2}{3}}{2} = 14.129 \quad \text{and} \quad s = 3.759$$

The 99% confidence interval is

$$\bar{x} \pm t_{.005} \frac{s}{\sqrt{n}} \Rightarrow 77.763 \pm 9.925 \frac{3.759}{\sqrt{3}} \Rightarrow 77.763 \pm 21.540$$

or $56.223 < \mu < 99.303$.

10.77 Using the formulas given in Chapter 2 or your scientific calculator, calculate

$$\bar{x} = \frac{\sum x_i}{n} = \frac{42.6}{10} = 4.26$$

$$s^2 = \frac{\sum x_i^2 - \frac{(\sum x_i)^2}{n}}{n-1} = \frac{190.46 - \frac{(42.6)^2}{10}}{9} = .998 \quad \text{and} \quad s = .999$$

The 95% confidence interval is then

$$\bar{x} \pm t_{.025} \frac{s}{\sqrt{n}} \Rightarrow 4.26 \pm 2.262 \frac{.999}{\sqrt{10}} \Rightarrow 4.26 \pm .715$$

or $3.545 < \mu < 4.975$.

10.79 The student may use the rounded values for $\bar{x}$ and s given in the display, or he may wish to calculate $\bar{x}$ and s and use the more exact calculations for the confidence intervals. The calculations are shown below.

a $\quad \bar{x} = \dfrac{\sum x_i}{n} = \dfrac{1845}{10} = 184.5 \qquad\qquad s^2 = \dfrac{\sum x_i^2 - \frac{(\sum x_i)^2}{n}}{n-1} = \dfrac{344,567 - \frac{(1845)^2}{10}}{9} = 462.7222$

$s = 21.511$ and the 95% confidence interval is

$$\bar{x} \pm t_{.025} \frac{s}{\sqrt{n}} \Rightarrow 1.845 \pm 2.262 \frac{21.511}{\sqrt{10}} \Rightarrow 184.5 \pm 15.4$$

or $169.1 < \mu < 199.9$.

b $\quad \bar{x} = \dfrac{\sum x_i}{n} = \dfrac{730}{10} = 73.0 \qquad\qquad s^2 = \dfrac{\sum x_i^2 - \frac{(\sum x_i)^2}{n}}{n-1} = \dfrac{53514 - \frac{(730)^2}{10}}{9} = 24.8889$

$s = 4.989$ and the 95% confidence interval is

$$\bar{x} \pm t_{.025} \frac{s}{\sqrt{n}} \Rightarrow 73.0 \pm 2.262 \frac{4.989}{\sqrt{10}} \Rightarrow 73.0 \pm 3.57$$

or $69.43 < \mu < 76.57$.

c $\quad \bar{x} = \dfrac{\sum x_i}{n} = \dfrac{25.42}{10} = 2.542 \qquad\qquad s^2 = \dfrac{\sum x_i^2 - \frac{(\sum x_i)^2}{n}}{n-1} = \dfrac{65.8398 - \frac{(25.42)^2}{10}}{9} = .13579556$

$s = .3685$ and the 95% confidence interval is

$$\bar{x} \pm t_{.025} \frac{s}{\sqrt{n}} \Rightarrow 2.54 \pm 2.262 \frac{.3685}{\sqrt{10}} \Rightarrow 2.54 \pm .26$$

or $2.28 < \mu < 2.80$.

128

d No. The relationship between the confidence intervals is not the same as the relationship between the original measurements.

10.81 A paired-difference analysis must be used. The hypothesis of interest is

$$H_0 : \mu_1 - \mu_2 = 0 \quad \text{or} \quad H_0 : \mu_d = 0$$
$$H_a : \mu_1 - \mu_2 < 0 \quad \text{or} \quad H_a : \mu_d < 0$$

The table of differences is presented below. Use your calculator to find $\bar{d}$ and s_d,

d_i	-3	-3	2	-1	-1	1	-3

Calculate $\bar{d} = -1.1429$, $s_d = 2.0354$, and the test statistic is

$$t = \frac{\bar{d} - \mu_d}{s_d / \sqrt{n}} = \frac{-1.1429 - 0}{\dfrac{2.0354}{\sqrt{7}}} = -1.49$$

The one-tailed rejection region with $df = n-1 = 6$ is $t < -1.943$ and H_0 is not rejected. There is insufficient evidence to indeicat that the mean reaction time is greater after consuming alcohol.

10.83 **a** The hypothesis to be tested is

$$H_0 : \mu_1 - \mu_2 = 0 \qquad H_a : \mu_1 - \mu_2 \neq 0$$

The following information is available:

$$\bar{x}_1 = 27.2 \qquad s_1^2 = 16.36 \quad \text{and} \quad n_1 = 10$$
$$\bar{x}_2 = 33.5 \qquad s_2^2 = 18.92 \quad \text{and} \quad n_2 = 10$$

Since the ratio of the variances is less than 3, you can use the pooled t test. The pooled estimator of σ^2 is calculated as

$$s^2 = \frac{(n_1 - 1)s_1^2 + (n_2 - 1)s_2^2}{n_1 + n_2 - 2} = \frac{9(16.36) + 9(18.92)}{10 + 10 - 2} = 17.64$$

and the test statistic is

$$t = \frac{(\bar{x}_1 - \bar{x}_2) - 0}{\sqrt{s^2 \left(\dfrac{1}{n_1} + \dfrac{1}{n_2}\right)}} = \frac{27.2 - 33.5}{\sqrt{17.64 \left(\dfrac{1}{10} + \dfrac{1}{10}\right)}} = -3.354$$

Critical value approach: The rejection region is two-tailed, based on $df = 18$ degrees of freedom. With $\alpha = .05$, from Table 4, the rejection region is $|t| > t_{.025} = 2.101$ and H_0 is rejected. There is sufficient evidence to indicate a difference in the mean absorption rates between the two drugs.

b **p-value approach:** The p-value is $2P(t > 3.354)$ for a two-tailed test with 18 degrees of freedom. Since $t = 3.354$ exceeds the largest tabled value, $t_{.005} = 2.878$, we have

$$p\text{-value} < 2(.005) = .01$$

Since the p-value is less than $\alpha = .05$, H_0 can be rejected at the 5% level of significance, confirming the results of part **a**.

c A 95% confidence interval for $(\mu_1 - \mu_2)$ is given as

$$(\bar{x}_1 - \bar{x}_2) \pm t_{.025} \sqrt{s^2 \left(\frac{1}{n_1} + \frac{1}{n_2}\right)}$$

$$(27.2 - 33.5) \pm 2.101 \sqrt{17.64 \left(\frac{1}{10} + \frac{1}{10}\right)}$$

$$-6.3 \pm 3.946 \quad \text{or} \quad -10.246 < (\mu_1 - \mu_2) < -2.354$$

129

Since the confidence interval does not contain the value $\mu_1 - \mu_2 = 0$, you can conclude that there is a difference in two population means. This confirms the conclusion in part **a**.

10.85   **a**   To check for equality of variances, we can use the rule of thumb or the formal F-test. First, calculate

$$s_1^2 = \frac{22{,}921{,}516 - \dfrac{15070^2}{10}}{9} = 23{,}447.3333 \quad \text{and} \quad s_2^2 = \frac{11{,}963{,}753 - \dfrac{10913^2}{10}}{9} = 6044.0111$$

The ratio

$$F = \frac{\text{larger } s^2}{\text{smaller } s^2} = \frac{23447.3333}{6044.0111} = 3.88$$

is greater than 3, which is an indication that the variances are unequal. Using a two-sided F-test with $df_1 = df_2 = 9$, the p-value for the test is bounded as

$$.025 < \frac{1}{2}(p\text{-value}) < .05$$
$$.05 < p\text{-value} < .10$$

Since the p-value is close to .05, we choose to assume that the variances are not equal, and use Satterthwaite's approximation.

b   The hypothesis of interest is $H_0 : \mu_1 - \mu_2 = 300$ versus $H_a : \mu_1 - \mu_2 > 300$ for the two independent samples of male and female pheasants. Calculate

$$t = \frac{(\bar{x}_1 - \bar{x}_2) - 0}{\sqrt{\dfrac{s_1^2}{n_1} + \dfrac{s_2^2}{n_2}}} = \frac{1507 - 1091.3 - 300}{\sqrt{\dfrac{23447.3333}{10} + \dfrac{6044.0111}{10}}} = 2.13$$

which has a t distribution with $df = \dfrac{\left(\dfrac{s_1^2}{n_1} + \dfrac{s_2^2}{n_2}\right)^2}{\dfrac{\left(\dfrac{s_1^2}{n_1}\right)^2}{n_1 - 1} + \dfrac{\left(\dfrac{s_2^2}{n_2}\right)^2}{n_2 - 1}} = 13.35 \approx 13$

A one-tailed rejection region is then $t > t_{.05} = 1.771$ and H_0 is rejected. There is sufficient evidence to indicate that the average weight of male pheasants exceeds that of females by more than 300 grams.

10.87   **a**   Use the computing formulas or your scientific calculator to calculate

$$\bar{x} = \frac{\sum x_i}{n} = \frac{.2753}{10} = .02753 \qquad s^2 = \frac{.00758155 - \dfrac{(.2753)^2}{10}}{9} = .000000282$$

$s = .00053135$ and the 99% confidence interval is

$$\bar{x} \pm t_{.005}\frac{s}{\sqrt{n}} \;\Rightarrow\; .02753 \pm 3.25\frac{.00053135}{\sqrt{10}} \;\Rightarrow\; .02753 \pm .00055$$

or $.02698 < \mu < .02808$.

b   Intervals constructed using this procedure will enclose μ 99% of the time in repeated sampling. Hence, we are fairly certain that this particular interval encloses μ.

c   The sample must be randomly selected, or at least behave as a random sample from the population of interest. For the chemist performing the analysis, this means that he or she must be certain that there is no unknown factor which is affecting the measurements, thus causing a biased sample rather than a representative sample.

10.89   **a**   The hypothesis of interest is

$$H_0 : \sigma_1^2 = \sigma_2^2 \quad \text{versus} \quad H_a : \sigma_1^2 \neq \sigma_2^2$$

The test statistic is

$$F = \frac{s_1^2}{s_2^2} = \frac{3.1}{1.4} = 2.21 \;.$$

The two-sided rejection region with $df_1 = 24$ and $df_2 = 24$ and $\alpha = .05$ is $F > F_{.025} = 2.27$ and the null hypothesis is not rejected. There is insufficient evidence to indicate a difference in the precision of the two machines.

b   The 95% confidence interval for the ratio of the two variances is

$$\frac{s_1^2}{s_2^2} \frac{1}{F_{df_1, df_2}} < \frac{\sigma_1^2}{\sigma_2^2} < \frac{s_1^2}{s_2^2} F_{df_2, df_1}$$

$$\frac{3.1}{1.4} \left(\frac{1}{2.27} \right) < \frac{\sigma_1^2}{\sigma_2^2} < \frac{3.1}{1.4} (2.27) \quad \text{or} \quad .975 < \frac{\sigma_1^2}{\sigma_2^2} < 5.03$$

Since the possible values for σ_1^2 / σ_2^2 includes the value 1, it is possible that there is no difference in the precision of the two machines. This confirms the results of part **a**.

10.91   A paired-difference analysis is used. The differences are shown below.

$$47, 44, 38, 46, 37, 56, 35, 34$$

Calculate

$$\bar{d} = \frac{\sum d_i}{n} = \frac{337}{8} = 42.125 \qquad s_d^2 = \frac{\sum d_i^2 - \frac{(\sum d_i)^2}{n}}{n-1} = \frac{14591 - \frac{(337)^2}{8}}{7} = 56.4107 \quad \text{and} \quad s_d = 7.5107$$

and the 95% confidence interval for $\mu_1 - \mu_2 = \mu_d$ is

$$\bar{d} \pm t_{.025} \frac{s_d}{\sqrt{n}} \;\Rightarrow\; 42.125 \pm 2.365 \frac{7.5107}{\sqrt{8}} \;\Rightarrow\; 42.125 \pm 6.280$$

or $35.845 < (\mu_1 - \mu_2) < 48.405$. The mean weight loss is somewhere between 36 and 48 pounds. We must assume that the 8 subjects represent a random sample and that the population of differences has a normal distribution.

10.93   **a**   Because of the mounded shape of the data shown in the stem and leaf plot, you can be assured that the assumption of normality is valid.

b   Using the summary statistics given in the printout, the 99% confidence interval is

$$\bar{x} \pm t_{.005} \frac{s}{\sqrt{n}} \;\Rightarrow\; 5.3953 \pm 2.756 \frac{.5462}{\sqrt{30}} \;\Rightarrow\; 5.3953 \pm .2748$$

or $5.1205 < \mu < 5.6701$.

10.95   Note that this experiment has been designed differently from the experiment performed in Exercise 10.94. That is, each of the eight people was subjected to both stimuli (in random order). A paired-difference analysis is used to test the hypothesis

$$H_0 : \mu_1 - \mu_2 = 0 \qquad H_a : \mu_1 - \mu_2 \neq 0$$

and the table of differences, along with the calculation of $\bar{d}$ and s_d^2, is presented below.

Person	1	2	3	4	5	6	7	8	Total
d_i	−1	−1	−2	1	−1	−1	0	−1	−6

131

$$\bar{d} = \frac{\sum d_i}{n} = \frac{-6}{8} = -.75 \qquad s_d^2 = \frac{\sum d_i^2 - \dfrac{\left(\sum d_i\right)^2}{n}}{n-1} = \frac{10 - \dfrac{(-6)^2}{8}}{7} = .78571 \qquad \text{and} \qquad s_d = .88641$$

and the test statistic is

$$t = \frac{\bar{d} - \mu_d}{s_d / \sqrt{n}} = \frac{-.75 - 0}{\dfrac{.88641}{\sqrt{8}}} = -2.39$$

The rejection region with $\alpha = .05$ and $df = n-1 = 7$ is $|t| > t_{.025} = 2.365$, and H_0 is rejected. There is a significant difference between the two stimuli.

10.97   **a**   The analysis is identical to that used in previous exercises. To test $H_0 : \mu_1 - \mu_2 = 0$ versus $H_a : \mu_1 - \mu_2 \neq 0$, the test statistic is

$$t = \frac{(\bar{x}_1 - \bar{x}_2) - 0}{\sqrt{s^2\left(\dfrac{1}{n_1} + \dfrac{1}{n_2}\right)}} = 9.5641$$

with $p\text{-value} = .0000$. Since this p-value is very small, H_0 is rejected. There is evidence of a difference in the means.
b   Since there is a difference in the mean strengths for the two kinds of material, the strongest material (A) should be used (all other factors being equal).

10.99   In order to use the F statistic to test the hypothesis concerning the equality of two population variances, we must assume that independent random samples have been drawn form two normal populations.

10.101   The hypothesis of interest is

$$H_0 : \sigma_A^2 = \sigma_B^2 \quad \text{versus} \quad H_a : \sigma_A^2 < \sigma_B^2$$

and the test statistic is

$$F = \frac{s_B^2}{s_A^2} = \frac{.065}{.027} = 2.407 \; .$$

The rejection region (one-tailed) will be determined by a critical value of F based on $df_1 = 9$ and $df_2 = 29$ degrees of freedom, with area .05 to its right. That is, from Table 6, you will reject H_0 if $F > 2.22$. The observed value of F falls in the rejection region, and we conclude that $\sigma_A^2 < \sigma_B^2$.

10.103   To test the hypothesis $H_0 : \mu = 16$ versus $H_a : \mu < 16$, the test statistic is

$$t = \frac{\bar{x} - \mu_0}{s / \sqrt{n}} = \frac{15.7 - 16}{.5 / \sqrt{9}} = -1.8$$

The p-value with 8 degrees of freedom is bounded as $.05 < p\text{-value} < .10$. Hence, the null hypothesis is not rejected. There is insufficient evidence to indicate that the mean weight is less than claimed.

10.105   **a**   Calculate

$$\bar{x}_1 = \frac{41.6}{7} = 5.943 \quad \text{and} \quad \bar{x}_2 = \frac{112.9}{7} = 16.129$$

$$s_1^2 = \frac{255.96 - \dfrac{(41.6)^2}{7}}{6} = 1.45619 \quad \text{and} \quad s_2^2 = \frac{1832.11 - \dfrac{(112.9)^2}{7}}{6} = 1.86571$$

Since the ratio of the variances is less than 3, you can use the pooled estimator of σ^2 calculated as

$$s^2 = \frac{(n_1 - 1)s_1^2 + (n_2 - 1)s_2^2}{n_1 + n_2 - 2} = \frac{6(1.45619) + 6(1.86571)}{12} = 1.66095$$

A 90% confidence interval for $(\mu_1 - \mu_2)$ is given as

$$(\bar{x}_1 - \bar{x}_2) \pm t_{.05} \sqrt{s^2 \left(\frac{1}{n_1} + \frac{1}{n_2} \right)}$$

$$(5.943 - 16.129) \pm 1.782 \sqrt{1.66095 \left(\frac{1}{7} + \frac{1}{7} \right)}$$

$$-10.186 \pm 1.228 \quad \text{or} \quad -11.414 < (\mu_1 - \mu_2) < -8.958$$

b Since the value $\mu_1 - \mu_2 = 0$ is not in the interval, it is unlikely that $\mu_1 = \mu_2$. Therefore, we conclude that there is evidence of a difference in the average amount of oil required to produce these two crops.

10.107 a In this exercise, the hypothesis of interest is

$$H_0 : \sigma_1^2 = \sigma_2^2 \quad \text{versus} \quad H_a : \sigma_1^2 \neq \sigma_2^2$$

and the test statistic is $\quad F = \frac{s_1^2}{s_2^2} = \frac{2.96}{1.54} = 1.922$.

The critical values of F for various levels of α are needed in order to find the approximate p-value. The critical values with $df_1 = 14$ and $df_2 = 14$ are not found in Table 6. However, they can be approximated as $F_{15,14}$ and are shown in the following table.

α	.10	.05	.025	.01	.005
F_α	2.01	2.46	2.95	3.66	4.25

Hence,

$$p\text{-value} = 2P(F > 1.922) > 2(.10) = .20$$

This is too large to reject H_0. There is insufficient evidence to indicate a difference in the variabilities of the closing prices of the two stocks.

b From the table in part **a**, $F_{df_1, df_2} = F_{df_2, df_1} \approx 4.25$. The 99% confidence interval for σ_1^2 / σ_2^2 is

$$\frac{s_1^2}{s_2^2} \frac{1}{F_{df_1, df_2}} < \frac{\sigma_1^2}{\sigma_2^2} < \frac{s_1^2}{s_2^2} F_{df_2, df_1}$$

$$1.922 \left(\frac{1}{4.25} \right) < \frac{\sigma_1^2}{\sigma_2^2} < 1.922(4.25) \quad \text{or} \quad .452 < \frac{\sigma_1^2}{\sigma_2^2} < 8.1685$$

Intervals constructed using this procedure will enclose the ratio σ_1^2 / σ_2^2 99% of the time in repeated sampling. Hence, we are fairly confident that the interval, .452 to 8.1685, encloses σ_1^2 / σ_2^2.

10.109 A paired-difference test is used. To test $H_0 : \mu_1 - \mu_2 = 0$ versus $H_a : \mu_1 - \mu_2 > 0$, where μ_1 is the mean before the safety program and μ_2 is the mean after the program, calculate the differences:

$$7, 6, -1, 5, 6, 1$$

Then $\bar{d} = \frac{\sum d_i}{n} = \frac{24}{6} = 4 \qquad s_d^2 = \frac{148 - \frac{(24)^2}{6}}{5} = 10.4 \quad \text{and} \quad s_d = 3.2249$

and the test statistic is

$$t = \frac{\bar{d} - \mu_d}{s_d / \sqrt{n}} = \frac{4 - 0}{\frac{3.2249}{\sqrt{6}}} = 3.038$$

133

For a one-tailed test with $df = 5$, the rejection region with $\alpha = .01$ is $t > t_{.01} = 3.365$, and H_0 is not rejected. There is insufficient evidence to indicate that the safety program was effective in reducing lost-time accidents.

10.111 Indexing $\chi^2_{.05}$ and $\chi^2_{.95}$ with $n - 1 = 19$ degrees of freedom in Table 5 yields

$$\chi^2_{.05} = 30.1435 \quad \text{and} \quad \chi^2_{.95} = 10.117$$

and the 90% confidence interval is

$$\frac{(n-1)s^2}{\chi^2_{.05}} < \sigma^2 < \frac{(n-1)s^2}{\chi^2_{.95}} \qquad \text{or} \quad 24.582 < \sigma^2 < 73.243$$
$$\frac{19(39)}{30.1435} < \sigma^2 < \frac{19(39)}{10.117}$$

10.113 a The hypothesis of interest is
$$H_0 : \sigma^2 = .03 \qquad H_a : \sigma^2 > .03$$

and the test statistic is

$$\chi^2 = \frac{(n-1)s^2}{\sigma^2_0} = \frac{14(.053)}{.03} = 24.73$$

The one-tailed rejection region with $\alpha = .05$ and $n - 1 = 14$ degrees of freedom is $\chi^2 > \chi^2_{.05} = 23.6848$, and H_0 is rejected. There is sufficient evidence to indicate that σ^2 is greater than .03.

b Indexing $\chi^2_{.025}$ and $\chi^2_{.975}$ with $n - 1 = 14$ degrees of freedom in Table 5 yields

$$\chi^2_{.025} = 26.1190 \quad \text{and} \quad \chi^2_{.975} = 5.62872$$

and the confidence interval is

$$\frac{14(.053)}{26.1190} < \sigma^2 < \frac{14(.053)}{5.62872} \qquad \text{or} \quad .0284 < \sigma^2 < .1318$$

10.115 Calculate $\bar{x} = \dfrac{\sum x_i}{n} = \dfrac{104.9}{25} = 4.196$

$$s^2 = \frac{\sum x_i^2 - \dfrac{\left(\sum x_i\right)^2}{n}}{n-1} = \frac{454.81 - \dfrac{(104.9)^2}{25}}{24} = .6104 \quad \text{and} \quad s = .7813$$

The 95% confidence interval based on $df = 24$ is

$$\bar{x} \pm t_{.025}\frac{s}{\sqrt{n}} \quad \Rightarrow \quad 4.196 \pm 2.064\frac{.7813}{\sqrt{25}} \quad \Rightarrow \quad 4.196 \pm .323$$

or $3.873 < \mu < 4.519$. Intervals constructed in this manner enclose the true value of μ 95% of the time. Hence, we are fairly certain that this interval contains the true value of μ.

10.117 a The hypothesis of interest is
$$H_0 : \sigma_1^2 = \sigma_2^2 \quad \text{versus} \quad H_a : \sigma_1^2 \neq \sigma_2^2$$

and the test statistic is $\quad F = \dfrac{s_1^2}{s_2^2} = \dfrac{22^2}{20^2} = 1.21$.

The upper portion of the two-tailed rejection region with $\alpha = .05$ is $F > F_{19,19} \approx F_{20,19} = 2.51$ and H_0 is not rejected. There is insufficient evidence to indicate that the population variances are different.

b The hypothesis to be tested is

$$H_0 : \mu_1 - \mu_2 = 0 \qquad H_a : \mu_1 - \mu_2 \neq 0$$

Based on the results of part **a**, you can use the pooled t test. The pooled estimator of σ^2 is calculated as

$$s^2 = \frac{(n_1 - 1)s_1^2 + (n_2 - 1)s_2^2}{n_1 + n_2 - 2} = \frac{19(22^2) + 19(20^2)}{38} = 442$$

and the test statistic is

$$t = \frac{(\bar{x}_1 - \bar{x}_2) - 0}{\sqrt{s^2\left(\dfrac{1}{n_1} + \dfrac{1}{n_2}\right)}} = \frac{78 - 67}{\sqrt{442\left(\dfrac{1}{20} + \dfrac{1}{20}\right)}} = 1.65$$

The rejection region with $\alpha = .05$ and $df = 20 + 20 - 2 = 38$ (approximated with $df = 29$) is $|t| > 2.045$ and H_0 is not rejected. There is insufficient evidence to indicate a difference in the two populaton means.

11: The Analysis of Variance

11.1   In comparing 6 populations, there are $k-1$ degrees of freedom for treatments and $n = 6(10) = 60$. The ANOVA table is shown below.

Source	df
Treatments	5
Error	54
Total	59

11.3   Refer to Exercise 11.2, where the analysis of variance table was calculated as shown below.

Source	df	SS	MS	F
Treatments	5	5.2	1.04	3.467
Error	54	16.2	0.30	
Total	59	21.4		

a $\bar{x}_1 \pm t_{.025}\sqrt{\dfrac{MSE}{n_1}} \Rightarrow 3.07 \pm 1.96\sqrt{\dfrac{.3}{10}} \Rightarrow 3.07 \pm .339$

or $2.731 < \mu_1 < 3.409$.

b $(\bar{x}_1 - \bar{x}_2) \pm t_{.025}\sqrt{MSE\left(\dfrac{1}{n_1} + \dfrac{1}{n_2}\right)}$

$(3.07 - 2.52) \pm 1.96\sqrt{0.3\left(\dfrac{2}{10}\right)}$

$.55 \pm .480$ or $.07 < \mu_1 - \mu_2 < 1.03$

11.5   **a**   Refer to Exercise 11.4. The given sums of squares are inserted and missing entries found by subtraction. The mean squares are found as $MS = SS/df$.

Source	df	SS	MS	F
Treatments	3	339.8	113.267	16.98
Error	20	133.4	6.67	
Total	23	473.2		

b   The F statistic, $F = MST/MSE$, has $df_1 = 3$ and $df_2 = 20$ degrees of freedom.

c   With $\alpha = .05$ and degrees of freedom from **b**, H_0 is rejected if $F > F_{.05} = 3.10$.

d   Since $F = 16.98$ falls in the rejection region, the null hypothesis is rejected. There is a difference among the means.

e   The critical values of F with $df_1 = 3$ and $df_2 = 20$ (Table 6) for bounding the p-value for this one-tailed test are shown below.

α	.10	.05	.025	.01	.005
F_α	2.38	3.10	3.86	4.94	5.82

Since the observed value $F = 16.98$ is greater than $F_{.005}$, p-value $< .005$, and H_0 is rejected as in part **d**.

11.7   The following preliminary calculations are necessary:

$$T_1 = 14 \quad T_2 = 19 \quad T_3 = 5 \quad G = 38$$

a $CM = \dfrac{\left(\sum x_{ij}\right)^2}{n} = \dfrac{(38)^2}{14} = 103.142857$

$$\text{Total SS} = \sum x_{ij}^2 - CM = 3^2 + 2^2 + \cdots + 2^2 + 1^2 - CM = 130 - 103.142857 = 26.8571$$

b $\quad SST = \sum \dfrac{T_i^2}{n_i} - CM = \dfrac{14^2}{5} + \dfrac{19^2}{5} + \dfrac{5^2}{4} - CM = 117.65 - 103.142857 = 14.5071$

and $MST = \dfrac{SST}{k-1} = \dfrac{14.5071}{2} = 7.2536$.

c By subtraction, $SSE = \text{Total SS} - SST = 26.8571 - 14.5071 = 12.3500$ and the degrees of freedom, by subtraction, are $13 - 2 = 11$. Then

$$MSE = \dfrac{SSE}{11} = \dfrac{12.3500}{11} = 1.1227$$

d The information obtained in parts **a-c** is consolidated in an ANOVA table.

Source	df	SS	MS
Treatments	2	14.5071	7.2536
Error	11	12.3500	1.1227
Total	13	26.8571	

e The hypothesis to be tested is

$$H_0 : \mu_1 = \mu_2 = \mu_3 \quad \text{versus} \quad H_a : \text{at least one pair of means are different}$$

f The rejection region for the test statistic $F = \dfrac{MST}{MSE} = \dfrac{7.2536}{1.1227} = 6.46$ is based on an F-distribution with 2 and 11 degrees of freedom. The critical values of F for bounding the p-value for this one-tailed test are shown below.

α	.10	.05	.025	.01	.005
F_α	2.86	3.98	5.26	7.21	8.91

Since the observed value $F = 6.46$ is between $F_{.01}$ and $F_{.025}$,

$$.01 < p\text{-value} < .025$$

and H_0 is rejected at the 5% level of significance. There is a difference among the means.

11.9   **a** The 90% confidence interval for μ_1 is

$$\overline{x}_1 \pm t_{.05} \sqrt{\dfrac{MSE}{n_1}} \ \Rightarrow\ 2.8 \pm 1.796 \sqrt{\dfrac{1.1227}{5}} \ \Rightarrow\ 2.8 \pm .85$$

or $1.95 < \mu_1 < 3.65$.

b The 90% confidence interval for $\mu_1 - \mu_3$ is

$$\left(\overline{x}_1 - \overline{x}_3\right) \pm t_{.05} \sqrt{MSE\left(\dfrac{1}{n_1} + \dfrac{1}{n_3}\right)}$$

$$\left(2.8 - 1.25\right) \pm 1.796 \sqrt{1.1227\left(\dfrac{1}{5} + \dfrac{1}{4}\right)}$$

$$1.55 \pm 1.28 \quad \text{or} \quad .27 < \mu_1 - \mu_3 < 2.83$$

and the F test to detect a difference in mean student response is

$$F = \dfrac{MST}{MSE} = 5.15.$$

The rejection region with $\alpha = .05$ and 2 and 8 df is $F > 4.46$ and H_0 is rejected. There is a significant difference in mean response due to the three different methods.

11.11   **a** The 95% confidence interval for μ_A is

$$\overline{x}_A \pm t_{.025} \sqrt{\dfrac{MSE}{n_A}} \ \Rightarrow\ 76 \pm 2.306 \sqrt{\dfrac{62.333}{5}} \ \Rightarrow\ 76 \pm 8.142$$

or $67.86 < \mu_A < 84.14$.

b The 95% confidence interval for μ_B is

$$\bar{x}_B \pm t_{.025} \sqrt{\frac{\text{MSE}}{n_B}} \quad \Rightarrow \quad 66.33 \pm 2.306 \sqrt{\frac{62.333}{3}} \quad \Rightarrow \quad 66.33 \pm 10.51$$

or $55.82 < \mu_B < 76.84$.

c The 95% confidence interval for $\mu_A - \mu_B$ is

$$\left(\bar{x}_A - \bar{x}_B\right) \pm t_{.025} \sqrt{\text{MSE}\left(\frac{1}{n_A} + \frac{1}{n_B}\right)}$$

$$\left(76 - 66.33\right) \pm 2.306 \sqrt{62.333\left(\frac{1}{5} + \frac{1}{3}\right)}$$

$$9.667 \pm 13.296 \quad \text{or} \quad -3.629 < \mu_A - \mu_B < 22.963$$

d Note that these three confidence intervals cannot be jointly valid because all three employ the same value of $s = \sqrt{\text{MSE}}$ and are dependent.

11.13 **a** We would be reasonably confident that the data satisfied the normality assumption because each measurement represents the average of 10 continuous measurements. The Central Limit Theorem assures us that this mean will be approximately normally distributed.

b We have a completely randomized design with four treatments, each containing 6 measurements. The analysis of variance table is given in the *Minitab* printout. The F test is

$$F = \frac{\text{MST}}{\text{MSE}} = \frac{6.580}{.115} = 57.38$$

with p-value = .000 (in the column marked "P"). Since the p-value is very small (less than .01), H_0 is rejected. There is a significant difference in the mean leaf length among the four locations with $P < .01$ or even $P < .001$.

c The hypothesis to be tested is $H_0 : \mu_1 = \mu_4$ versus $H_a : \mu_1 \neq \mu_4$ and the test statistic is

$$t = \frac{\bar{x}_1 - \bar{x}_4}{\sqrt{\text{MSE}\left(\frac{1}{n_1} + \frac{1}{n_4}\right)}} = \frac{6.0167 - 3.65}{\sqrt{.115\left(\frac{1}{6} + \frac{1}{6}\right)}} = 12.09$$

The p-value with $df = 20$ is $2P\left(t > 12.09\right)$ is bounded (using Table 4) as

$$p\text{-value} < 2\left(.005\right) = .01$$

and the null hypothesis is reject. We conclude that there is a difference between the means.

d The 99% confidence interval for $\mu_1 - \mu_4$ is

$$\left(\bar{x}_1 - \bar{x}_4\right) \pm t_{.005} \sqrt{\text{MSE}\left(\frac{1}{n_1} + \frac{1}{n_4}\right)}$$

$$\left(6.0167 - 3.65\right) \pm 2.845 \sqrt{.115\left(\frac{1}{6} + \frac{1}{6}\right)}$$

$$2.367 \pm .557 \quad \text{or} \quad 1.810 < \mu_1 - \mu_4 < 2.924$$

e When conducting the t tests, remember that the stated confidence coefficients are based on random sampling. If you looked at the data and only compared the largest and smallest sample means, the randomness assumption would be disturbed.

11.15 The design is completely randomized with 3 treatments and 5 replications per treatment. The *Minitab* printout on the next page shows the analysis of variance for this experiment.

138

One-way ANOVA: Calcium versus Method

```
Source   DF         SS         MS         F       P
Method    2   0.0000041  0.0000021    16.38   0.000
Error    12   0.0000015  0.0000001
Total    14   0.0000056

S = 0.0003545   R-Sq = 73.19%   R-Sq(adj) = 68.72%

                        Individual 95% CIs For Mean Based on
                                     Pooled StDev
Level  N     Mean      StDev   -+---------+---------+---------+--------
1      5   0.027620  0.000421                      (------*------)
2      5   0.026780  0.000396   (------*------)
3      5   0.028040  0.000207                            (------*------)
                              -+---------+---------+---------+--------
                             0.02650   0.02700   0.02750   0.02800
Pooled StDev = 0.000354
```

The test statistic, $F = 16.38$ with p-value $= .000$ indicates the results are highly significant; there is a difference in the mean calcium contents for the three methods. All assumptions appear to have been satisfied.

11.17   **a**   The design is a completely randomized design (four independent samples).

     b   The following preliminary calculations are necessary:

$$T_1 = 1311 \quad T_2 = 1174 \quad T_3 = 1258 \quad T_4 = 1343 \quad G = 5086$$

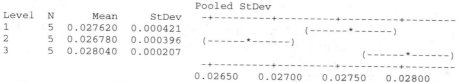

$$CM = \frac{\left(\sum x_{ij}\right)^2}{n} = \frac{(5086)^2}{20} = 1,293,369.8 \ \text{Total SS} = \sum x_{ij}^2 - CM = 1,297,302 - CM = 3932.2$$

$$SST = \sum \frac{T_i^2}{n_i} - CM = \frac{1311^2}{5} + \frac{1174^2}{5} + \frac{1258^2}{5} + \frac{1343^2}{5} - CM = 3272.2$$

Calculate $MS = SS/df$ and consolidate the information in an ANOVA table.

Source	df	SS	MS
Treatments	3	3272.2	1090.7333
Error	16	660	41.25
Total	19	3932.2	

    c   The hypothesis to be tested is

$$H_0 : \mu_1 = \mu_2 = \mu_3 = \mu_4 \quad \text{versus} \quad H_a : \text{at least one pair of means are different}$$

and the F test to detect a difference in average prices is

$$F = \frac{MST}{MSE} = 26.44 \ .$$

The rejection region with $\alpha = .05$ and 3 and 16 df is approximately $F > 3.24$ and H_0 is rejected. [Alternatively, we could bound the p-value using Table 6 as p-value $< .005$.] There is enough evidence to indicate a difference in the average prices for the four states.

11.19   Sample means must be independent and based upon samples of equal size.

11.21   Use Tables 11(a) and 11(b).

    a $\omega = q_{.05}(4,12) \dfrac{s}{\sqrt{5}} = 4.20 \dfrac{s}{\sqrt{5}} = 1.878s$

    b $\omega = q_{.01}(6,12) \dfrac{s}{\sqrt{8}} = 6.10 \dfrac{s}{\sqrt{8}} = 2.1567s$

11.23 With $k = 4$, $df = 20$, $n_t = 6$,

$$\omega = q_{.01}(4,20)\frac{\sqrt{\text{MSE}}}{\sqrt{n_t}} = 5.02\sqrt{\frac{.115}{6}} = .69$$

The ranked means are shown below.

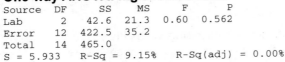

11.25 The design is completely randomized with 3 treatments and 5 replications per treatment. The *Minitab* printout below shows the analysis of variance for this experiment.

One-way ANOVA: mg/dl versus Lab

```
Source   DF     SS     MS     F      P
Lab       2   42.6   21.3   0.60   0.562
Error    12  422.5   35.2
Total    14  465.0
S = 5.933    R-Sq = 9.15%    R-Sq(adj) = 0.00%

                        Individual 95% CIs For Mean Based on
                        Pooled StDev
Level  N    Mean   StDev   --+---------+---------+---------+-------
1      5  108.86    7.47                  (-------------*-------------)
2      5  105.04    6.01      (---------------*-------------)
3      5  105.60    3.70       (-------------*-------------)
                            --+---------+---------+---------+-------
                            100.0     104.0     108.0     112.0

Pooled StDev = 5.93

Tukey 95% Simultaneous Confidence Intervals
All Pairwise Comparisons among Levels of Lab
Individual confidence level = 97.94%
Lab = 1 subtracted from:
Lab   Lower   Center  Upper    +---------+---------+---------+---------
2   -13.824   -3.820  6.184    (---------------*-------------)
3   -13.264   -3.260  6.744       (-------------*--------------)
                                +---------+---------+---------+---------
                               -14.0      -7.0      0.0       7.0


Lab = 2 subtracted from:
Lab   Lower   Center  Upper    +---------+---------+---------+---------
3    -9.444    0.560  10.564        (--------------*-------------)
                                +---------+---------+---------+---------
                               -14.0      -7.0      0.0       7.0
```

a The analysis of variance F test for $H_0: \mu_1 = \mu_2 = \mu_3$ is $F = .60$ with *p*-value $= .562$. The results are not significant and H_0 is not rejected. There is insufficient evidence to indicate a difference in the treatment means.

b Since the treatment means are not significantly different, there is no need to use Tukey's test to search for the pairwise differences. Notice that all three intervals generated by *Minitab* contain zero, indicating that the pairs cannot be judged different.

11.27 **a** The following preliminary calculations are necessary:

$$T_1 = 6080 \quad T_2 = 6530 \quad T_3 = 5320 \quad G = 17{,}930$$

$$\text{CM} = \frac{\left(\sum x_{ij}\right)^2}{n} = \frac{(17{,}930)^2}{30} = 10{,}716{,}163.333333$$

$$\text{Total SS} = \sum x_{ij}^2 - \text{CM} = 11{,}016{,}900 - \text{CM} = 300{,}736.666667$$

$$\text{SST} = \sum \frac{T_i^2}{n_i} - \text{CM} = \frac{6080^2}{10} + \frac{6530^2}{10} + \frac{5320^2}{10} - \text{CM} = 74{,}806.666667$$

Calculate $MS = SS/df$ and consolidate the information in an ANOVA table.

Source	df	SS	MS
Treatments	2	74,806.666667	37,403.333333
Error	27	225,930.000000	8367.777778
Total	29	300,736.666667	

The hypothesis to be tested is

$$H_0 : \mu_1 = \mu_2 = \mu_3 \quad \text{versus} \quad H_a : \text{at least one pair of means are different}$$

and the F test to detect a difference in average scores is

$$F = \frac{MST}{MSE} = 4.47 .$$

The rejection region with $\alpha = .05$ and 2 and 27 df is $F > 3.35$ and H_0 is rejected. There is evidence of a difference in the average scores for the three graduate programs.

b The 95% confidence interval for $\mu_1 - \mu_2$ is

$$\left(\overline{x}_{LS} - \overline{x}_{PS} \right) \pm t_{.025} \sqrt{ MSE \left(\frac{1}{n_{LS}} + \frac{1}{n_{PS}} \right) }$$

$$\left(\frac{6080}{10} - \frac{6530}{10} \right) \pm 2.052 \sqrt{ 8367.777778 \left(\frac{1}{10} + \frac{1}{10} \right) }$$

$$-45 \pm 83.946 \quad \text{or} \quad -128.946 < \mu_{LS} - \mu_{PS} < 38.946$$

c With $k = 3$, $df = 27$, $n_i = 10$,

$$\omega = q_{.05} (3, 27) \frac{\sqrt{MSE}}{\sqrt{n_i}} \approx 3.53 \sqrt{ \frac{8367.777778}{10} } = 102.113$$

The ranked means are shown below.

$$
\begin{array}{ccc}
532 & 608 & 653 \\
\overline{x}_{SS} & \overline{x}_{LS} & \overline{x}_{PS}
\end{array}
$$

There is no significant difference between Social Sciences and Life Sciences, or between Life Sciences and Physical Sciences; but the average scores for Social Sciences and Physical Sciences are different from each other.

11.29 Refer to Exercise 11.28. The given sums of squares are inserted and missing entries found by subtraction. The mean squares are found as $MS = SS/df$.

Source	df	SS	MS	F
Treatments	2	11.4	5.70	4.01
Blocks	5	17.1	3.42	2.41
Error	10	14.2	1.42	
Total	17	42.7		

11.31 The 95% confidence interval for $\mu_A - \mu_B$ is then

$$\left(\overline{x}_A - \overline{x}_B \right) \pm t_{.025} \sqrt{ MSE \left(\frac{2}{b} \right) }$$

$$(21.9 - 24.2) \pm 2.228 \sqrt{ 1.42 \left(\frac{2}{6} \right) }$$

$$-2.3 \pm 1.533 \quad \text{or} \quad -3.833 < \mu_A - \mu_B < -.767$$

141

11.33 Use *Minitab* or *MS Excel* to obtain an ANOVA printout, or use the following calculations:

$$CM = \frac{\left(\sum x_{ij}\right)^2}{n} = \frac{(113)^2}{12} = 1064.08333$$

Total SS $= \sum x_{ij}^2 - CM = 6^2 + 10^2 + \cdots + 14^2 - CM = 1213 - CM = 148.91667$

$$SST = \sum \frac{T_j^2}{3} - CM = \frac{22^2 + 34^2 + 27^2 + 30^2}{3} - CM = 25.58333$$

$$SSB = \sum \frac{B_i^2}{4} - CM = \frac{33^2 + 25^2 + 55^2}{4} - CM = 120.66667 \text{ and}$$

SSE = Total SS $-$ SST $-$ SSB $= 2.6667$

Calculate MS $=$ SS$/df$ and consolidate the information in an ANOVA table.

Source	df	SS	MS	F
Treatments	3	25.5833	8.5278	19.19
Blocks	2	120.6667	60.3333	135.75
Error	6	2.6667	0.4444	
Total	11	148.9167		

a To test the difference among treatment means, the test statistic is

$$F = \frac{MST}{MSE} = \frac{8.528}{.4444} = 19.19$$

and the rejection region with $\alpha = .05$ and 3 and 6 df is $F > 4.76$. There is a significant difference among the treatment means.

b To test the difference among block means, the test statistic is

$$F = \frac{MSB}{MSE} = \frac{60.3333}{.4444} = 135.75$$

and the rejection region with $\alpha = .05$ and 2 and 6 df is $F > 5.14$. There is a significant difference among the block means.

c With $k = 4$, $df = 6$, $n_i = 3$,

$$\omega = q_{.01}(4,6)\frac{\sqrt{MSE}}{\sqrt{n_i}} = 7.03\sqrt{\frac{.4444}{3}} = 2.71$$

The ranked means are shown below.

7.33	9.00	10.00	11.33
$\bar{x}_1$	$\bar{x}_3$	$\bar{x}_4$	$\bar{x}_2$

d The 95% confidence interval is

$$\left(\bar{x}_A - \bar{x}_B\right) \pm t_{.025}\sqrt{MSE\left(\frac{2}{b}\right)}$$

$$\left(7.333 - 11.333\right) \pm 2.447\sqrt{.4444\left(\frac{2}{3}\right)}$$

$$-4 \pm 1.332 \quad \text{or} \quad -5.332 < \mu_A - \mu_B < -2.668$$

e Since there is a significant difference among the block means, blocking has been effective. The variation due to block differences can be isolated using the randomized block design.

11.35 **a** By subtraction, the degrees of freedom for blocks is $b - 1 = 34 - 28 = 6$. Hence, there are $b = 7$ blocks.
b There are always $b = 7$ observations in a treatment total.
c There are $k = 4 + 1 = 5$ observations in a block total.

d

Source	df	SS	MS	F
Treatments	4	14.2	3.55	9.68
Blocks	6	18.9	3.15	8.59
Error	24	8.8	0.3667	
Total	34	41.9		

e To test the difference among treatment means, the test statistic is

$$F = \frac{MST}{MSE} = \frac{3.55}{.3667} = 9.68$$

and the rejection region with $\alpha = .05$ and 4 and 24 df is $F > 2.78$. There is a significant difference among the treatment means.

f To test the difference among block means, the test statistic is

$$F = \frac{MSB}{MSE} = \frac{3.15}{.3667} = 8.59$$

and the rejection region with $\alpha = .05$ and 6 and 24 df is $F > 2.51$. There is a significant difference among the block means.

d To determine where the treatment differences lie, use Tukey's test with

$$\omega = q_{.05}(3,6)\frac{\sqrt{MSE}}{\sqrt{n_t}} = 4.34\sqrt{\frac{.224167}{4}} = 1.027$$

The ranked means are shown below.

$$
\begin{array}{ccc}
26.525 & 27.05 & 27.725 \\
\overline{x}_A & \overline{x}_C & \overline{x}_B
\end{array}
$$

Only gasoline formulations A and B are significantly different from each other.

11.37 Similar to previous exercises. The *Minitab* printout for this randomized block experiment is shown below.

Two-way ANOVA: Measurements versus Blocks, Chemicals

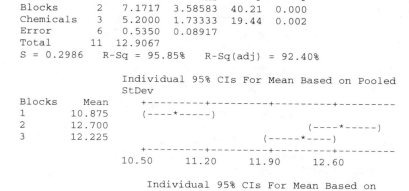

```
Source      DF      SS       MS       F       P
Blocks       2   7.1717   3.58583   40.21   0.000
Chemicals    3   5.2000   1.73333   19.44   0.002
Error        6   0.5350   0.08917
Total       11  12.9067
S = 0.2986    R-Sq = 95.85%    R-Sq(adj) = 92.40%


                  Individual 95% CIs For Mean Based on Pooled
                  StDev
Blocks    Mean    +---------+---------+---------+---------
1        10.875   (----*-----)
2        12.700                             (----*-----)
3        12.225                     (-----*----)
                  +---------+---------+---------+---------
                 10.50    11.20     11.90     12.60


                   Individual 95% CIs For Mean Based on
                   Pooled StDev
Chemicals   Mean   ------+---------+---------+---------+---
1         11.4000  (-----*-----)
2         12.3333               (-----*-----)
3         11.2000  (-----*-----)
4         12.8000                        (-----*-----)
                   ------+---------+---------+---------+---
                      11.20     11.90     12.60     13.30
```

Both the treatment and block means are significantly different. Since the four chemicals represent the treatments in this experiment, Tukey's test can be used to determine where the differences lie:

$$\omega = q_{.05}(4,6)\frac{\sqrt{MSE}}{\sqrt{n_t}} = 4.90\sqrt{\frac{.08917}{3}} = .845$$

The ranked means are shown below.

11.20	11.40	12.33	12.80
$\overline{x}_3$	$\overline{x}_1$	$\overline{x}_2$	$\overline{x}_4$

The chemical falls into two significantly different groups – A and C versus B and D.

11.39 The factor of interest is "soil preparation", and the blocking factor is "locations". A randomized block design is used and the analysis of variance table can be obtained using the computer printout.

a The F statistic to detect a difference due to soil preparations is

$$F = \frac{MST}{MSE} = 10.06$$

with p-value $= .012$. The null hypothesis can be rejected at the 5% level of significance; there is a significant difference among the treatment means.

b The F statistic to detect a difference due to locations is

$$F = \frac{MSB}{MSE} = 10.88$$

with p-value $= .008$. The null hypothesis can be rejected at the 1% level of significance; there is a highly significant difference among the block means.

c Tukey's test can be used to determine where the differences lie:

$$\omega = q_{.05}(3,6)\frac{\sqrt{MSE}}{\sqrt{n_t}} = 4.34\sqrt{\frac{1.8889}{4}} = 2.98$$

The ranked means are shown below.

12.0	12.5	16.0
$\overline{x}_3$	$\overline{x}_1$	$\overline{x}_2$

Preparations 2 and 3 are the only two treatments that can be declared significantly different.

d The 95% confidence interval is

$$(\overline{x}_B - \overline{x}_A) \pm t_{.025}\sqrt{MSE\left(\frac{2}{b}\right)}$$

$$(16.5 - 12.5) \pm 2.447\sqrt{1.89\left(\frac{2}{4}\right)}$$

$$3.5 \pm 2.38 \quad \text{or} \quad 1.12 < \mu_B - \mu_A < 5.88$$

11.41 A randomized block design has been used with "estimators" as treatments and "construction job" as the block factor. The analysis of variance table is found in the *Minitab* printout below.

Two-way ANOVA: Cost versus Estimator, Job

Source	DF	SS	MS	F	P
Estimator	2	10.8617	5.4308	7.20	0.025
Job	3	37.6073	12.5358	16.61	0.003
Error	6	4.5283	0.7547		
Total	11	52.9973			

S = 0.8687   R-Sq = 91.46%   R-Sq(adj) = 84.34%

```
                    Individual 95% CIs For Mean Based on
                    Pooled StDev
Estimator    Mean   -------+---------+---------+---------+--
A          32.6125  (--------*--------)
B          34.8875                   (--------*--------)
C          34.1875             (--------*--------)
                    -------+---------+---------+---------+--
                      32.4      33.6      34.8      36.0
```

144

Both treatments and blocks are significant. The treatment means can be further compared using Tukey's test with

$$\omega = q_{.05}(3,6)\frac{\sqrt{MSE}}{\sqrt{n_t}} = 4.34\sqrt{\frac{.7547}{4}} = 1.885$$

The ranked means are shown below.

$$\begin{array}{ccc} 32.6125 & 34.1875 & 34.8875 \\ \overline{x}_A & \overline{x}_C & \overline{x}_B \end{array}$$

Estimators A and B show a significant difference in average costs.

11.43 **a** A randomized block design has been used with "stores" as treatments and "items" as the block factor.
b The F statistic to detect a difference in mean prices for the five stores is

$$F = \frac{MST}{MSE} = 25.53$$

with p-value $= .000$. The null hypothesis of no difference should be rejected; there is evidence of a significant difference in average prices among the stores.
c The F statistic to detect a difference in mean prices for the 8 items is

$$F = \frac{MSB}{MSE} = 29.99$$

with p-value $= .000$. The null hypothesis of no difference should be rejected; there is evidence of a significant difference in average prices from item to item. That is, blocking has been effective.

11.45 **a-b** There are $4 \times 5 = 20$ treatments and $4 \times 5 \times 3 = 60$ total observations.
c In a factorial experiment, variation due to the interaction $A \times B$ is isolated from SSE. The sources of variation and associated degrees of freedom are given on the next page.

Source	df
A	3
B	4
$A \times B$	12
Error	40
Total	59

11.47 Refer to Exercise 11.46, where we constructed the ANOVA table shown below.

Source	df	SS	MS	F
A	2	5.3	2.6500	1.30
B	3	9.1	3.0333	1.49
$A \times B$	6	4.8	0.8000	0.39
Error	12	24.5	2.0417	
Total	23	43.7		

The 95% confidence interval is

$$(\overline{x}_1 - \overline{x}_2) \pm t_{.025}\sqrt{MSE\left(\frac{2}{r}\right)}$$

$$(8.3 - 6.3) \pm 2.179\sqrt{2.0417\left(\frac{2}{2}\right)}$$

$$2.0 \pm 3.11 \quad \text{or} \quad -1.11 < \mu_1 - \mu_2 < 5.11$$

11.49 **a** Based on the fact that the mean response for the two levels of factor B behaves very differently depending on the level of factor A under investigation, there is a strong interaction present between factors A and B.

    b The test statistic for interaction is $F = \text{MS}(AB)/\text{MSE} = 37.85$ with p-value $= .000$ from the *Minitab* printout. There is evidence of a significant interaction. That is, the effect of factor A depends upon the level of factor B at which A is measured.

    c In light of this type of interaction, the main effect means (averaged over the levels of the other factor) differ only slightly. Hence, a test of the main-effect terms produces a non-significant result.

    d No. A significant interaction indicates that the effect of one factor depends upon the level of the other. Each factor-level combination should be investigated individually.

    e Answers will vary.

11.51 **a** The total number of participants was sixty, twenty in each of three categories. Hence, the total degrees of freedom is fifty-nine. Factor T was run at two levels, factor A at three levels, resulting in the given degrees of freedom.

    b $F = \dfrac{\text{MST}}{\text{MSE}} = \dfrac{103.7009}{28.3015} = 3.66$ $F = \dfrac{\text{MSA}}{\text{MSE}} = \dfrac{760.5889}{28.3015} = 26.87$

        $F = \dfrac{\text{MS}(TA)}{\text{MSE}} = \dfrac{124.9905}{28.3015} = 4.42$

    c Since interaction is significant, the main effects need not be tested individually. Attention should be focused on the individual cell means.

    d The tabled values for the approximate *df* are shown below.

α	.10	.05	.025	.01	.005
$F_\alpha(1,60)$	2.79	4.00	5.29	7.08	8.49
$F_\alpha(2,60)$	2.39	3.15	3.93	4.98	5.79

For factor T, $.05 < p\text{-value} < .10$. For factor A, $p\text{-value} < .005$, and for the interaction $A \times T$, $.01 < p\text{-value} < .025$.

11.53 **a** The design is a 2×4 factorial experiment with $r = 5$ replications. There are two factors, Gender and School, one at two levels and one at four levels.

    b The analysis of variance table can be found using a computer printout or the following calculations:

	Schools				
Gender	1	2	3	4	Total
Male	2919	3257	3330	2461	11967
Female	3082	3629	3344	2410	12465
Total	6001	6886	6674	4871	24432

$\text{CM} = \dfrac{24432^2}{40} = 14923065.6$ $\text{Total SS} = 15281392 - \text{CM} = 358326.4$

$\text{SSG} = \dfrac{11967^2 + 12465^2}{20} - \text{CM} = 6200.1$

$\text{SS}(Sc) = \dfrac{6001^2 + 6886^2 + 6674^2 + 4871^2}{10} - \text{CM} = 246725.8$

$\text{SS}(G \times Sc) = \dfrac{2919^2 + 3257^2 + \cdots + 2410^2}{5} - \text{SSG} - \text{SS}(Sc) - \text{CM} = 10574.9$

Source	df	SS	MS	F
G	1	6200.1	6200.100	2.09
Sc	3	246725.8	82241.933	27.75
G×Sc	3	10574.9	3524.967	1.19
Error	32	94825.6	2963.300	
Total	39	358326.4		

146

c   The test statistic is $F = \text{MS(GSc)}/\text{MSE} = 1.19$ and the rejection region is $F > 2.92$ (with $\alpha = .05$). Alternately, you can bound the p-value $> .10$. Hence, H_0 is not rejected. There is insufficient evidence to indicate interaction between gender and schools.

d   You can see in the interaction plot that there is a small difference between the average scores for male and female students at schools 1 and 2, but no difference to speak of at the other two schools. The interaction is not significant.

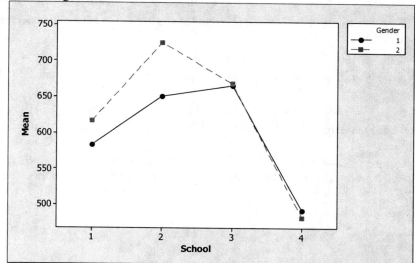

e   The test statistic for testing gender is $F = 2.09$ with $F_{.05} = 4.17$ (or p-value $> .10$). The test statistic for schools is $F = 27.75$ with $F_{.05} = 2.92$ (or p-value $< .005$). There is a significant effect due to schools. Using Tukey's method of paired comparisons with $\alpha = .01$, calculate

$$\omega = q_{.01}(4,32)\frac{\sqrt{\text{MSE}}}{\sqrt{n_t}} = 4.80\sqrt{\frac{2963.3}{10}} = 82.63$$

The ranked means are shown below.

$$\begin{array}{cccc} 487.1 & 600.1 & 667.4 & 688.6 \\ \bar{x}_4 & \bar{x}_1 & \bar{x}_3 & \bar{x}_2 \end{array}$$

11.55   **a**   The analysis of variance table can be found using a computer printout or the following calculations:

	Training (A)		
Situation (B)	*Trained*	*Not Trained*	*Total*
Standard	334	185	185
Emergency	296	177	473
Total	630	362	992

$$CM = \frac{992^2}{16} = 61504 \qquad \text{Total SS} = 66640 - CM = 5136$$

$$SSA = \frac{630^2 + 362^2}{8} - CM = 65993 - 61504 = 4489$$

$$SSB = \frac{519^2 + 473^2}{8} - CM = 132.25$$

$$SS(A \times B) = \frac{334^2 + 296^2 + \cdots + 117^2}{4} - SSA - SSB - CM = 56.25$$

Source	df	SS	MS	F
A	1	4489	4489	117.49
B	1	132.25	132.25	3.46
A×B	1	56.25	56.25	1.47
Error	12	458.5	38.2083	
Total	15	5136		

b The test statistic is $F = MS(A \times B)/MSE = 1.47$ and the rejection region is $F > 4.75$ (with $\alpha = .05$). Alternately, you can bound the p-value $> .10$. Hence, H_0 is not rejected. The interaction term is not significant.

c The test statistic is $F = MSB/MSE = 3.46$ and the rejection region is $F > 4.75$ (with $\alpha = .05$). Alternately, you can bound the $.05 < p$-value $< .10$. Hence, H_0 is not rejected. Factor B (Situation) is not significant.

d The test statistic is $F = MSA/MSE = 117.49$ and the rejection region is $F > 4.75$ (with $\alpha = .05$). Alternately, you can bound the p-value $< .005$. Hence, H_0 is rejected. Factor A (Training) is highly significant.

e The interaction plot is shown on the next page. The response is much higher for the supervisors who have been trained. You can see very little change in the response for the two different situations (standard or emergency). The parallel lines indicate that there is no interaction between the two factors.

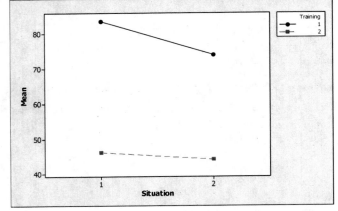

11.57 The intervals provided in the *Minitab* printout allow you to declare a difference between a pair of means only when both endpoints have the same sign. Significant differences are observed between treatments A and C, B and C, C and E and D and E. The ranked means are shown below.

E	A	B	D	C
.48	.62	.67	.92	1.07

11.59 The objective is to determine whether or not mean reaction time differs for the five stimuli. The four people used in the experiment act as blocks, in an attempt to isolate the variation from person to person. A randomized block design is used, and the analysis of variance table is given in the *Excel* printout.

a The F statistic to detect a difference due to stimuli is

$$F = \frac{MST}{MSE} = 27.776$$

with p-value $= .000$. There is a significant difference in the effect of the five stimuli.

b The treatment means can be further compared using Tukey's test with

$$\omega = q_{.05}(5,12)\frac{\sqrt{MSE}}{\sqrt{n_t}} = 4.51\sqrt{\frac{.00708}{4}} = .190$$

The ranked means are shown on the next page.

	E	A	B	D	C
	.525	.7	.8	1.025	1.05

c The F test for blocks produces $F = 6.588$ with p-value $= .007$. The block differences are significant; blocking has been effective.

11.61 Answers will vary from student to student. A completely randomized design has been used. The analysis of variance table is shown in the printout.

One-way ANOVA: 1, 2, 3, 4
```
Source   DF     SS      MS      F      P
Factor    3   1385.8   461.9   9.84   0.000
Error    23   1079.4    46.9
Total    26   2465.2
S = 6.851   R-Sq = 56.21%   R-Sq(adj) = 50.50%

                        Individual 95% CIs For Mean Based on
                        Pooled StDev
Level   N    Mean   StDev   ----+---------+---------+---------+-----
1       6   80.333  8.595              (------*-------)
2       8   91.875  4.912                         (-----*-----)
3       5   80.400  4.930            (-------*------)
4       8   73.500  7.964   (-----*-----)
                        ----+---------+---------+---------+-----
                        72.0      80.0      88.0      96.0

Pooled StDev = 6.851

Tukey 95% Simultaneous Confidence Intervals
All Pairwise Comparisons

Individual confidence level = 98.90%
1 subtracted from:

      Lower   Center   Upper   ---------+---------+---------+---------+
2     1.313   11.542  21.771                     (------*------)
3   -11.402    0.067  11.536            (-------*-------)
4   -17.062   -6.833   3.396          (-----*------)
                        ---------+---------+---------+---------+
                            -15        0        15        30

2 subtracted from:
      Lower   Center   Upper   ---------+---------+---------+---------+
3   -22.273  -11.475  -0.677          (------*-------)
4   -27.845  -18.375  -8.905   (------*-----)
                        ---------+---------+---------+---------+
                            -15        0        15        30

3 subtracted from:
      Lower   Center   Upper   ---------+---------+---------+---------+
4   -17.698   -6.900   3.898          (------*-------)
                        ---------+---------+---------+---------+
                            -15        0        15        30
```
The student should recognize the significant difference in the mean responses for the four training programs, and should further investigate these differences using Tukey's test with ranked means shown below:

	4	1	3	2
	73.5	80.33	80.4	91.875

11.63 This is similar to previous exercises. The complete ANOVA table is shown below.

Source	df	SS	MS	F
A	1	1.14	1.14	6.51
B	2	2.58	1.29	7.37
A×B	2	0.49	0.245	1.40
Error	24	4.20	0.175	
Total	29	8.41		

149

a The test statistic is $F = MS(AB)/MSE = 1.40$ and the rejection region is $F > 3.40$. There is insufficient evidence to indicate an interaction.

b Using Table 6 with $df_1 = 2$ and $df_2 = 24$, the following critical values are obtained.

α	.10	.05	.025	.01	.005
F_α	2.54	3.40	4.32	5.61	6.66

The observed value of F is less than $F_{.10}$, so that p-value $> .10$.

c The test statistic for testing factor A is $F = 6.51$ with $F_{.05} = 4.26$. There is evidence that factor A affects the response.

d The test statistic for factor B is $F = 7.37$ with $F_{.05} = 3.40$. Factor B also affects the response.

11.65 **a** The experiment is an 2×3 factorial and a two-way analysis of variance is generated.

b Using the *Excel* printout given in the exercise, the F test for interaction is $F = .452$ with p-value $= .642$. There is insufficient evidence to suggest that the effect of temperature is different depending on the type of plant.

c The plot of treatment means for cotton and cucumber as a function of temperature is shown on the next page. The temperature appears to have a quadratic effect on the number of eggs laid in both cotton and cucumber. However, the treatment means are higher overall for the cucumber plants.

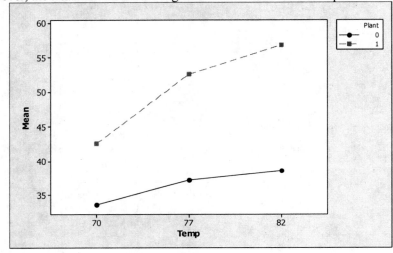

d The 95% confidence interval for $\mu_{Cotton} - \mu_{Cucumber}$ is

$$\left(\overline{x}_{Cotton} - \overline{x}_{Cucumber}\right) \pm t_{.025}\sqrt{MSE\left(\frac{1}{n_{Cotton}} + \frac{1}{n_{Cucumber}}\right)}$$

$$\left(\frac{547}{15} - \frac{760}{15}\right) \pm 2.064\sqrt{123.017\left(\frac{2}{15}\right)}$$

$$-14.2 \pm 8.36 \quad \text{or} \quad -22.56 < \mu_{Cotton} - \mu_{Cucumber} < -5.84$$

11.67 **a** The design is a randomized block design, with weeks representing blocks and stores as treatments. It can also be analyzed as the paired difference experiment discussed in Chapter 10.

b Use *Minitab* or *MS Excel* to obtain an ANOVA printout, or use the following calculations:

$$CM = \frac{\left(\sum x_{ij}\right)^2}{n} = \frac{(1988.89)^2}{8} = 494,460.429$$

$$\text{Total SS} = \sum x_{ij}^2 - CM = 495,350.089 - CM = 889.66$$

$$SS(\text{stores}) = \sum \frac{T_j^2}{b} - CM = \frac{(960.91)^2 + (1027.98)^2}{4} - CM = 562.298125$$

$$SS(\text{weeks}) = \sum \frac{B_i^2}{k} - CM = \frac{(510.29)^2 + (496.27)^2 + (487.02)^2 + (495.31)^2}{2} - CM = 139.70775 \text{ and}$$

$$SSE = \text{Total SS} - SST - SSB = 187.654125$$

Calculate $MS = SS/df$ and consolidate the information in an ANOVA table.

Source	df	SS	MS	F
Store	1	562.298	562.298	8.99
Week	3	139.708	46.569	0.74
Error	3	187.654	62.551	
Total	7	889.660		

c The F test for treatments is $F = 8.99$ with $.05 < p\text{-value} < .10$. The p-value is not small enough to allow rejection of H_0. There is insufficient evidence to indicate a significant difference in the average weekly totals for the two supermarket chains.

11.69 **a** This is a factorial experiment. The *Minitab* analysis of variance is shown below.

Two-way ANOVA: VO2 versus Gender, Activity
```
Source        DF      SS       MS       F       P
Gender         1  290.510  290.510  205.69  0.000
Activity       2  167.516   83.758   59.30  0.000
Interaction    2   21.501   10.750    7.61  0.004
Error         18   25.422    1.412
Total         23  504.950

S = 1.188   R-Sq = 94.97%   R-Sq(adj) = 93.57%
```

b The F-test for interaction is $F = 7.61$ with a p-value of .004. There is evidence of significant interaction between gender and levels of physical activity. The F-test for gender is $F = 205.69$ with a p-value of .000. There is sufficient evidence to indicate a difference in the average maximum oxygen uptake due to gender. The F-test for levels of physical activity is $F = 59.30$ with a p-value of .000. There is sufficient evidence to indicate a difference in the average maximum oxygen uptake due to levels of physical activity.

c Since the interaction was significant, Tukey's test is used to explore the differences in the six factor-level means with

$$\omega = q_{.05}(6,18) \frac{\sqrt{MSE}}{\sqrt{n_t}} \approx 4.49 \sqrt{\frac{1.412}{4}} = 2.67$$

The ranked means are shown below.

FL	FS	FM	ML	MS	MM
36.03	38.25	40.17	40.58	45.40	49.35

The males who have higher levels of physical activity have significantly higher average maximum oxygen uptakes.

11.71 **a** The design is a completely randomized design with three samples, each having a different number of measurements.

b Use the computing formulas in Section 11.5 or the *Minitab* printout on the next page.

One-way ANOVA: Iron versus Site

```
Source   DF       SS      MS       F      P
Site      2  132.277  66.139  126.85  0.000
Error    21   10.950   0.521
Total    23  143.227

S = 0.7221   R-Sq = 92.36%   R-Sq(adj) = 91.63%
```

The F test for treatments has a test statistic $F = 126.85$ with p-value = .000. The null hypothesis is rejected and we conclude that there is a significant difference in the average percentage of iron oxide at the three sites.

c The diagnostic plots are shown below. There appears to be no violation of the normality assumptions; the variances may be unequal, judging by the differing bar widths above and below the center line.

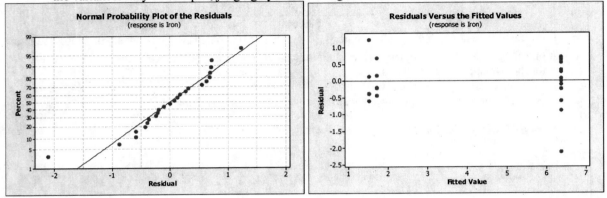

11.73 There is no evidence of non-normality, but there appears to be a difference in the variability within some of the factor-level combinations.

12: Linear Regression and Correlation

12.1   The line corresponding to the equation $y = 2x + 1$ can be graphed by locating the y values corresponding to $x = 0$, 1, and 2.

$$\text{When } x = 0, y = 2(0) + 1 = 1$$
$$\text{When } x = 1, \ y = 2(1) + 1 = 3$$
$$\text{When } x = 2, y = 2(2) + 1 = 5$$

The graph is shown below.

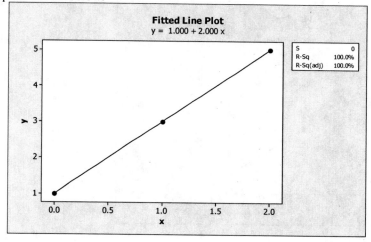

Note that the equation is in the form

$$y = \alpha + \beta x \, .$$

Thus, the slope of the line is $\beta = 2$ and the y-intercept is $\alpha = 1$.

12.3   If $\alpha = 3$ and $\beta = -1$, the straight line is $y = 3 - x$.

12.5   A deterministic mathematical model is a model in which the value of a response y is exactly predicted from values of the variables that affect the response. On the other hand, a probabilistic mathematical model is one that contains random elements with specific probability distributions. The value of the response y in this model is not exactly determined.

12.7   **a**   It is necessary to obtain a prediction equation relating y to x that provides the "best fit" to the data. The "best fitting" line is one which minimizes the sum of squares of the deviations of the observed y-values from the prediction equation. This line, called the "least squares" line, is denoted by

$$\hat{y} = a + bx \, .$$

The equations for calculating the quantities a and b are found in Section 12.2 of the text and involve the preliminary calculations:

$$\sum x_i = 21 \qquad \sum y_i = 24.3 \qquad \sum x_i y_i = 75.3$$
$$\sum x_i^2 = 91 \qquad \sum y_i^2 = 103.99 \qquad n = 6$$

Then

$$S_{xy} = \sum x_i y_i - \frac{\left(\sum x_i\right)\left(\sum y_i\right)}{n} = 75.3 - \frac{21(24.3)}{6} = 75.3 - 85.05 = -9.75$$

$$S_{xx} = \sum x_i^2 - \frac{(\sum x_i)^2}{n} = 91 - \frac{21^2}{6} = 17.5$$

$$b = \frac{S_{xy}}{S_{xx}} = \frac{-9.75}{17.5} = -0.55714 \quad \text{and} \quad a = \bar{y} - b\bar{x} = \frac{24.3}{6} - (-0.557)\left(\frac{21}{6}\right) = 6$$

and the least squares line is

$$\hat{y} = a + bx = 6 - 0.557x.$$

b   The graph of the least squares line and the six data points are shown below.

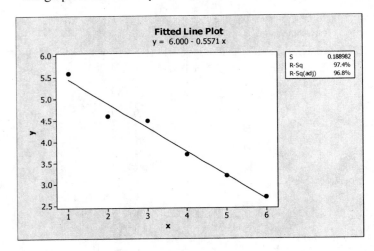

c   When $x = 3.5$, the value for y can be predicted using the least squares line as

$$\hat{y} = 6.00 - 0.55714(3.5) = 4.05$$

d   Using the additivity properties for the sums of sums of squares and degrees of freedom for an analysis of variance, and the fact that $MS = SS/df$, the completed ANOVA table is shown below.

```
Analysis of Variance
Source       DF      SS        MS
Regression    1    5.4321    5.4321
Error         4    0.1429    0.0357
Total         5    5.5750
```

12.9   **a**   The equations for calculating the quantities a and b are found using the preliminary calculations:

$$\sum x_i = 1490 \qquad \sum y_i = 1978 \qquad \sum x_i y_i = 653,830$$
$$\sum x_i^2 = 540,100 \qquad \sum y_i^2 = 827,504 \qquad n = 5$$

Then

$$S_{xy} = \sum x_i y_i - \frac{(\sum x_i)(\sum y_i)}{n} = 653,830 - \frac{1490(1978)}{5} = 64,386$$

$$S_{xx} = \sum x_i^2 - \frac{(\sum x_i)^2}{n} = 540,100 - \frac{1490^2}{5} = 96,080$$

$$b = \frac{S_{xy}}{S_{xx}} = \frac{64,386}{96,080} = 0.670129 \quad \text{and} \quad a = \bar{y} - b\bar{x} = 395.6 - 0.670129(298) = 195.902$$

and the least squares line is

$$\hat{y} = a + bx = 195.90 + 0.67x.$$

b   The graph of the least squares line and the six data points are shown below.

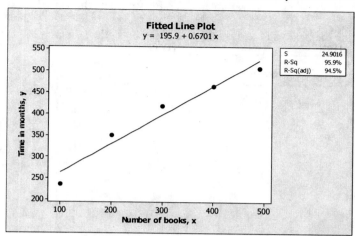

Fitted Line Plot
y = 195.9 + 0.6701 x

S	24.9016
R-Sq	95.9%
R-Sq(adj)	94.5%

c   Calculate Total $SS = S_{yy} = \sum y_i^2 - \dfrac{(\sum y_i)^2}{n} = 827,504 - \dfrac{(1978)^2}{5} = 45007.2$. Then

$$SSR = \dfrac{(S_{xy})^2}{S_{xx}} = \dfrac{64386^2}{96080} = 43146.9296$$

and $SSE = \text{Total SS} - SSR = S_{yy} - \dfrac{(S_{xy})^2}{S_{xx}} = 1860.2704$

The ANOVA table with 1 *df* for regression and $n - 2$ *df* for error is shown on the next page. Remember that the mean squares are calculated as $MS = SS/df$.

Source	*df*	SS	MS
Regression	1	43,146.9296	43,146.9296
Error	3	1860.2704	620.0901
Total	4	45,007.2000	

12.11   **a-b**   There are $n = 2(5) = 10$ pairs of observations in the experiment, so that the total number of degrees of freedom are $n - 1 = 9$.

    c   Using the additivity properties for the sums of sums of squares and degrees of freedom for an analysis of variance, and the fact that $MS = SS/df$, the completed ANOVA table is shown below.

Regression Analysis: y versus x

```
The regression equation is
y = 3.00 + 0.475 x

Predictor    Coef    SE Coef    T      P
Constant     3.000   2.127      1.41   0.196
x            0.4750  0.1253     3.79   0.005

S = 2.24165    R-Sq = 64.2%    R-Sq(adj) = 59.8%

Analysis of Variance
Source          DF      SS       MS       F       P
Regression      1       72.200   72.200   14.37   0.005
Residual Error  8       40.200   5.025
Total           9       112.400
```

d   From the computer printout the least squares line is $\hat{y} = a + bx = 3.00 + 0.475x$.

e   When $x = 10$, the value for y can be predicted using the least squares line as

$$\hat{y} = a + bx = 3.00 + 0.475(10) = 7.75.$$

155

12.13 **a** The API score might be partially explained by the percentage of students who are English Learners. That is, API would be the dependent or response variable, y, and EL would be the independent or explanatory variable, x.

b The scatterplot is shown on the next page. Notice the linear pattern in the points.

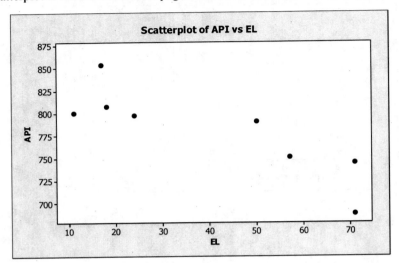

c Calculate

$$\sum x_i = 319 \qquad \sum y_i = 6236 \qquad \sum x_i y_i = 241,125$$

$$\sum x_i^2 = 17,141 \qquad \sum y_i^2 = 4,878,636 \qquad n = 8$$

Then

$$S_{xy} = \sum x_i y_i - \frac{(\sum x_i)(\sum y_i)}{n} = -7535.5$$

$$S_{xx} = \sum x_i^2 - \frac{(\sum x_i)^2}{n} = 4420.875$$

$$S_{yy} = \sum y_i^2 - \frac{(\sum y_i)^2}{n} = 17,674$$

Then

$$b = \frac{S_{xy}}{S_{xx}} = \frac{-7535.5}{4420.875} = -1.7045268 \quad \text{and} \quad a = \overline{y} - b\overline{x} = 779.5 - (-1.7045268)(39.875) = 847.468$$

and the least squares line is

$$\hat{y} = a + bx = 847.468 - 1.7045x.$$

The fitted line and the plotted points are shown below.

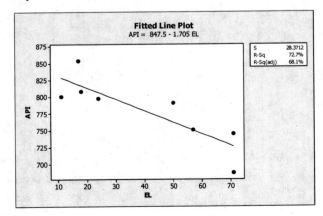

156

12.15 **a** The scatterplot generated by *Minitab* is shown below. The assumption of linearity is reasonable.

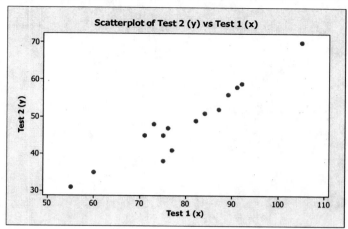

Scatterplot of Test 2 (y) vs Test 1 (x)

b Calculate

$$\sum x_i = 1192 \qquad \sum y_i = 725 \qquad \sum x_i y_i = 59,324$$

$$\sum x_i^2 = 96,990 \qquad \sum y_i^2 = 36,461 \qquad n = 15$$

Then

$$S_{xy} = \sum x_i y_i - \frac{(\sum x_i)(\sum y_i)}{n} = 1710.6667$$

$$S_{xx} = \sum x_i^2 - \frac{(\sum x_i)^2}{n} = 2265.7333$$

$$S_{yy} = \sum y_i^2 - \frac{(\sum y_i)^2}{n} = 1419.3333$$

$$b = \frac{S_{xy}}{S_{xx}} = \frac{1710.6667}{2265.7333} = .75502 \text{ and } a = \overline{y} - b\overline{x} = 48.3333 - (0.75502)(79.4667) = -11.665$$

(using full accuracy) and the least squares line is

$$\hat{y} = a + bx = -11.665 + 0.755x.$$

c When $x = 85$, the value for y can be predicted using the least squares line as

$$\hat{y} = a + bx = -11.665 + .755(85) = 52.51.$$

12.17 **a** The scatterplot is shown below. There is a positive linear relationship between arm span and height.

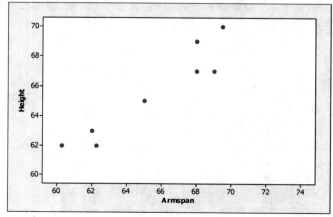

157

b   If arm span and height are roughly equal, the slope of the regression line should be approximately equal to 1.

c   Calculate $n = 8$; $\sum x_i = 524$; $\sum y_i = 525$; $\sum x_i^2 = 34413.375$; $\sum y_i^2 = 34521$; $\sum x_i y_i = 34462$.

Then

$$S_{xy} = \sum x_i y_i - \frac{(\sum x_i)(\sum y_i)}{n} = 74.5$$

$$S_{xx} = \sum x_i^2 - \frac{(\sum x_i)^2}{n} = 91.375$$

$$S_{yy} = \sum y_i^2 - \frac{(\sum y_i)^2}{n} = 67.875$$

$$b = \frac{S_{xy}}{S_{xx}} = \frac{74.5}{91.375} = .8153215 \quad \text{and} \quad a = \bar{y} - b\bar{x} = 65.625 - (0.8153215)(65.5) = 12.221$$

and the least squares line is

$$\hat{y} = a + bx = 12.221 + 0.815x.$$

The slope is quite close to the expected value of $\beta = 1$.

e   When $x = 62$, the value for y can be predicted using the least squares line as

$$\hat{y} = a + bx = 12.221 + .815(62) = 62.75.$$

12.19   **a**   The hypothesis to be tested is

$$H_0 : \beta = 0 \quad \text{versus} \quad H_a : \beta \neq 0$$

and the test statistic is a Student's t, calculated as

$$t = \frac{b - \beta_0}{\sqrt{MSE/S_{xx}}} = \frac{1.2 - 0}{\sqrt{0.533/10}} = 5.20$$

The critical value of t is based on $n - 2 = 3$ degrees of freedom and the rejection region for $\alpha = 0.05$ is $|t| > t_{.025} = 3.182$. Since the observed value of t falls in the rejection region, we reject H_0 and conclude that $\beta \neq 0$. That is, x is useful in the prediction of y.

b   From the ANOVA table in Exercise 12.6, calculate

$$F = \frac{MSR}{MSE} = \frac{14.4}{0.5333} = 27.00$$

which is the square of the t statistic from part **a**: $t^2 = (5.20)^2 = 27.0$.

c   The critical value of t from part **a** is $t_{.025} = 3.182$, while the critical value of F from part **b** with $df_1 = 1$ and $df_2 = 3$ is $F_{.05} = 10.13$. Notice that the relationship between the two critical values is

$$F = 10.13 = (3.182)^2 = t^2$$

12.21   **a**   The hypothesis to be tested is

$$H_0 : \beta = 0 \quad \text{versus} \quad H_a : \beta \neq 0$$

and the test statistic is

$$F = \frac{MSR}{MSE} = \frac{5.4321}{0.0357} = 152.10$$

with p-value $= 0.000$. Since the p-value is less than $\alpha = 0.01$, the null hypothesis is rejected. There is evidence to indicate that y and x are linearly related.

b   Use the formula for r^2 given in this section:

$$r^2 = \frac{SSR}{Total\ SS} = \frac{5.4321}{5.5750} = 0.974.$$

158

The coefficient of determination measures the proportion of the total variation in y that is accounted for using the independent variable x. That is, the total variation in y is reduced by 97.4% by using $\hat{y} = a + bx$ rather than $\bar{y}$ to predict the response y.

12.23   **a**   The equations for calculating the quantities a and b involve the following preliminary calculations:

$$\sum x_i = 797 \qquad\qquad \sum y_i = 169 \qquad\qquad \sum x_i y_i = 13,586$$
$$\sum x_i^2 = 64,063 \qquad\qquad \sum y_i^2 = 2887 \qquad\qquad n = 10$$

Then

$$S_{xy} = \sum x_i y_i - \frac{\left(\sum x_i\right)\left(\sum y_i\right)}{n} = 13,586 - \frac{797(169)}{10} = 116.7$$

$$S_{xx} = \sum x_i^2 - \frac{\left(\sum x_i\right)^2}{n} = 64,063 - \frac{797^2}{10} = 542.1$$

$$b = \frac{S_{xy}}{S_{xx}} = .215274 \quad\text{and}\quad a = \bar{y} - b\bar{x} = 16.9 - .215274(79.7) = -.2573$$

and the least squares line is

$$\hat{y} = a + bx = -.2573 + .2153x.$$

b   Calculate $\text{Total SS} = S_{yy} = \sum y_i^2 - \frac{\left(\sum y_i\right)^2}{n} = 2887 - \frac{(169)^2}{10} = 30.9.$ Then

$$\text{SSE} = S_{yy} - \frac{\left(S_{xy}\right)^2}{S_{xx}} = 30.9 - \frac{(116.7)^2}{542.1} = 5.777532$$

and $\text{MSE} = \dfrac{\text{SSE}}{n-2} = \dfrac{5.777532}{8} = 0.72219.$ The hypothesis to be tested is

$$H_0 : \beta = 0 \quad\text{versus}\quad H_a : \beta \neq 0$$

and the test statistic is

$$t = \frac{b - \beta_0}{\sqrt{\text{MSE}/S_{xx}}} \quad \frac{.2153 - 0}{\sqrt{0.72219/542.1}} = 5.90$$

The critical value of t is based on $n - 2 = 8$ degrees of freedom and the rejection region for $\alpha = 0.05$ is $|t| > t_{.025} = 2.306$, and H_0 is rejected. There is evidence at the 5% level to indicate that x and y are linearly related.

c   Calculate

$$r^2 = \frac{\left(S_{xy}\right)^2}{S_{xx}S_{yy}} = \frac{(116.7)^2}{(542.1)(30.9)} = 0.813$$

Then 81.3% of the total variation in y is accounted for by the independent variable x. That is, the total variation in y is reduced by 81.3% by using $\hat{y} = a + bx$ rather than $\bar{y}$ to predict the response y.

12.25   **a**   Refer to Exercise 12.9 to find

$$\text{Total SS} = S_{yy} = \sum y_i^2 - \frac{\left(\sum y_i\right)^2}{n} = 827,504 - \frac{(1978)^2}{5} = 45,007.2$$

$$S_{xx} = 96,080 \quad\text{and}\quad S_{xy} = 64,386 . \text{ Then}$$

$$\text{SSE} = S_{yy} - \frac{\left(S_{xy}\right)^2}{S_{xx}} = 45,007.2 - \frac{(64,386)^2}{96,080} = 1860.2704$$

and $\text{MSE} = \dfrac{\text{SSE}}{n-2} = \dfrac{1860.2704}{3} = 620.09013$. The hypothesis to be tested is

$$H_0 : \beta = 0 \quad \text{versus} \quad H_a : \beta \neq 0$$

and the test statistic is

$$t = \frac{b - \beta_0}{\sqrt{\text{MSE}/S_{xx}}} = \frac{0.670129 - 0}{\sqrt{620.09013/96,080}} = 8.34$$

The critical value of t is based on $n - 2 = 3$ degrees of freedom and the observed value $t = 8.34$ is larger than $t_{.005}$ so that p-value < 0.005. Hence, we reject H_0 and conclude that $\beta \neq 0$. That is, x is useful in the prediction of y.

b   Calculate

$$r^2 = \frac{S_{xy}^2}{S_{xx}S_{yy}} = \frac{64,386^2}{(96,080)(45,007.2)} = 0.959$$

The total variation has been reduced by 95.9% by using the linear model.

c   Refer to the plot in the solution to Exercise 12.9. The points show a curvilinear rather than a linear pattern. Although the fit as measured by r^2 is quite good, it may be that we have fit the wrong type of model to the data.

12.27   Refer to Exercise 12.18, where the least squares line was found to be
$$\hat{y} = a + bx = 2 - .875x.$$

and the ANOVA table was constructed as shown below.

Source	df	SS	MS
Regression	1	12.25	12.250
Error	3	0.25	.0833
Total	4	12.50	

a   The best estimate of σ^2 is MSE = .25/3 = .08333.

b   The hypothesis to be tested is
$$H_0 : \beta = 0 \quad \text{versus} \quad H_a : \beta \neq 0$$

and the test statistic is a Student's t, calculated as

$$t = \frac{b - \beta_0}{\sqrt{\text{MSE}/S_{xx}}} = \frac{-.875 - 0}{\sqrt{0.08333/16}} = -12.124$$

The critical value of t is based on $n - 2 = 3$ degrees of freedom and the rejection region for $\alpha = 0.05$ is $|t| > t_{.025} = 3.182$. Since the observed value of t falls in the rejection region, we reject H_0 and conclude that $\beta \neq 0$. That is, texture and storage temperature are linearly related.

c   Calculate

$$r^2 = \frac{\text{SSR}}{\text{Total SS}} = \frac{12.25}{12.5} = 0.98$$

d   The total variation has been reduced by 98% by using the linear model.

12.29   **a**   From the *Minitab* printout, R-sq = 75.7% or $r^2 = .757$. We can confirm this by calculating

$$r^2 = \frac{\text{SSR}}{\text{Total SS}} = \frac{3254}{4299} = 0.757$$

b   The total variation has been reduced by 75.7% by using the linear model.

12.31   Use a normal probability plot of the residuals. The residuals should approximate a straight line, sloping upward.

12.33   Use a plot of residuals versus fits. The plot should appear as a random scatter of points, free of any patterns.

12.35   The two plots behave as expected if the regression assumptions have been satisfied.

12.37   **a**   The random error ε must have a normal distribution with mean 0 and a common variance σ^2, independent of x.

       b   The best estimate of σ^2 is MSE = 58.1.

       c   The normal probability plot shows a slight deviation from normality in the tails of the distribution. The residual plot may have one unusual observation (upper left), but there do not appear to be any extreme violations of the regression assumptions.

12.39   **a**   In order to obtain an estimate for the expected value of y for a given value of x (or for a particular value of y), it would seem reasonable to use the prediction equation, $\hat{y} = a + bx$. Notice that x_p represents the given value of x for which we are estimating $E(y)$. The point estimator for $E(y)$ when $x = 1$ is

$$\hat{y} = 3 + 1.2(1) = 4.2$$

and the 90% confidence interval is

$$\hat{y} \pm t_{.05}\sqrt{\text{MSE}\left(\frac{1}{n} + \frac{(x_p - \bar{x})^2}{S_{xx}}\right)}$$

$$4.2 \pm 2.353\sqrt{(0.5333)\left(\frac{1}{5} + \frac{(1-0)^2}{10}\right)}$$

$$4.2 \pm 0.941$$

or $3.259 < E(y) < 5.141$.

       b   It is necessary to find a 90% prediction interval for y when $x = 1$. The interval used in predicting a particular value of y is

$$\hat{y} \pm t_{.05}\sqrt{\text{MSE}\left(1 + \frac{1}{n} + \frac{(x_p - \bar{x})^2}{S_{xx}}\right)}$$

$$4.2 \pm 2.353\sqrt{(0.5333)\left(1 + \frac{1}{5} + \frac{(1-0)^2}{10}\right)}$$

$$4.2 \pm 1.96$$

or $2.24 < y < 6.16$. We are 90% confident that the true value of y when $x = 1$ is in the above interval. Note that the above interval is much wider than the interval calculated for the expected value of y. The variability of predicting a particular value of y is greater than the variability of predicting the population mean for a particular value of x.

12.41   **a**   Use a computer program or the hand calculations shown below.

$$\sum x_i = 45 \qquad \sum y_i = 132 \qquad \sum x_i y_i = 411$$
$$\sum x_i^2 = 145 \qquad \sum y_i^2 = 1204 \qquad n = 15$$

161

Then

$$S_{xy} = \sum x_i y_i - \frac{\left(\sum x_i\right)\left(\sum y_i\right)}{n} = 411 - \frac{45(132)}{15} = 15$$

$$S_{xx} = \sum x_i^2 - \frac{\left(\sum x_i\right)^2}{n} = 145 - \frac{45^2}{15} = 10$$

$$S_{yy} = \sum y_i^2 - \frac{\left(\sum y_i\right)^2}{n} = 1204 - \frac{132^2}{15} = 42.4$$

$$b = \frac{S_{xy}}{S_{xx}} = \frac{15}{10} = 1.5 \quad \text{and} \quad a = \overline{y} - b\overline{x} = 8.8 - 1.5(3) = 4.3$$

and the least squares line is

$$\hat{y} = a + bx = 4.3 + 1.5x.$$

b The graph of the least squares line and the 15 data points are shown on the next page.

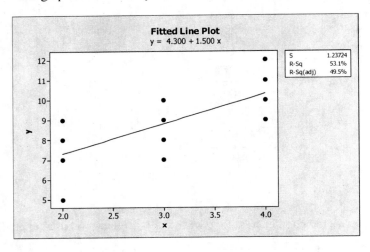

c Calculate

$$\text{SSE} = S_{yy} - \frac{\left(S_{xy}\right)^2}{S_{xx}} = 42.4 - \frac{15^2}{10} = 19.9$$

and $s^2 = \dfrac{\text{SSE}}{n-2} = 1.53$.

d To test $H_0 : \beta = 0, H_a : \beta \neq 0$, the test statistic is

$$t = \frac{b - \beta_0}{s/\sqrt{S_{xx}}} = \frac{1.5}{\sqrt{1.53/10}} = 3.83$$

The rejection region for $\alpha = 0.05$ is $|t| > t_{.025} = 2.160$ and we reject H_0. There is sufficient evidence to indicate that the independent variable x does help in predicting values of the dependent variable y.

e From Table 4, notice that $t = 3.834$ is larger than the largest tabulated value with $n - 2 = 13$ degrees of freedom $(t_{.005} = 3.012)$. Hence, the p-value for this two-tailed test is

$$2P(t > 3.834) < 2(0.005) = 0.01$$

f The diagnostic plots are shown below. There is no reason to doubt the validity of the regression assumptions.

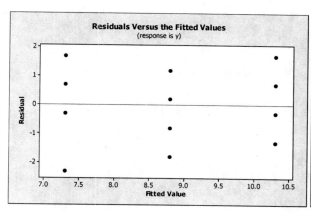

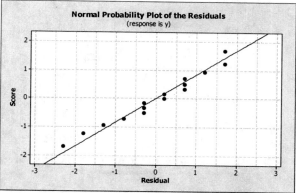

g   When $x = 3$, $\hat{y} = 4.3 + 1.5(3) = 8.8$ and the 95% confidence interval is

$$\hat{y} \pm t_{.025}\sqrt{\text{MSE}\left(\frac{1}{n} + \frac{(x_p - \bar{x})^2}{S_{xx}}\right)}$$

$$8.8 \pm 2.160\sqrt{1.53\left(\frac{1}{15} + \frac{(3-3)^2}{10}\right)}$$

$$8.8 \pm 0.69$$

or $8.11 < E(y) < 9.49$.

12.43   **a**   From the computer printout, the least squares line is $\hat{y} = a + bx = 251,206 + 27.406x$. The 99% confidence interval for β is

$$b \pm t_{.005} \times (\text{std error of } b) \Rightarrow 27.406 \pm 3.169(1.828) \Rightarrow 27.406 \pm 5.793$$

or $21.613 < \beta < 33.199$.

b   When $x = 2000$, $\hat{y} = 251,206 + 27.406(2000) = 306,018$ and the 95% confidence interval is given on the printout as

$$304,676 < E(y) < 307,360.$$

c   For each house in the sample, the price per square foot is calculated as $z_i = y_i / x_i$, and the results are shown below.

```
197.740   146.727   172.920   194.196   162.232   131.702   154.476
184.472   191.111   169.528   167.106   140.659
```

Then the average cost per square foot is

$$\bar{z} = \frac{\sum z_i}{n} = 167.739$$

This is not the same as $b = 27.406$ and should not be, since they are calculated in totally different ways.

d   When $x = 1780$, $\hat{y} = 299,989$ and the 95% prediction interval is given on the printout as

$$295,826 < y < 304,151$$

12.45   **a**   **a**   Use a computer program or the hand calculations shown below.

$$\sum x_i = 438 \qquad \sum y_i = 4620 \qquad \sum x_i y_i = 127,944$$

$$\sum x_i^2 = 12,294 \qquad \sum y_i^2 = 1,377,862 \qquad n = 16$$

Then

$$S_{xy} = \sum x_i y_i - \frac{(\sum x_i)(\sum y_i)}{n} = 1471.5$$

$$S_{xx} = \sum x_i^2 - \frac{(\sum x_i)^2}{n} = 303.75$$

$$S_{yy} = \sum y_i^2 - \frac{(\sum y_i)^2}{n} = 43,837$$

$$b = \frac{S_{xy}}{S_{xx}} = \frac{1471.5}{303.75} = 4.844\overline{4} \quad \text{and} \quad a = \overline{y} - b\overline{x} = 288.75 - 4.844\overline{4}(27.375) = 156.135$$

and the least squares line is
$$\hat{y} = 156.135 + 4.844x.$$

b The proportion of the total variation explained by regression is

$$r^2 = \frac{S_{xy}^2}{S_{xx}S_{yy}} = \frac{(1471.5)^2}{(303.75)(43,837)} = 0.163$$

c The diagnostic plots, generated by *Minitab* are shown below. The plots do not show any strong violation of assumptions.

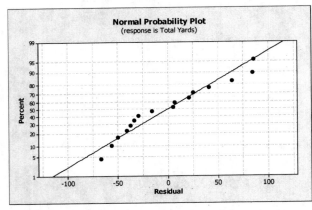

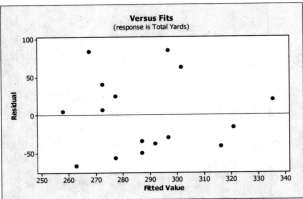

12.47 Note, however, that r^2 provides a meaningful measure of the strength of the linear relationship between two variables, y and x. It is the ratio of the reduction in the sum of squares of deviations obtained using the model, $y = \alpha + \beta x + \varepsilon$, to the sum of squares of deviations that would be obtained if the variable x were ignored. That is, r^2 measures the amount of variation that can be attributed to the variable x.

12.49 **a** $r = +1$
 b $r = -1$

12.51 **a** Refer to the figure below. The sample correlation coefficient will be negative.

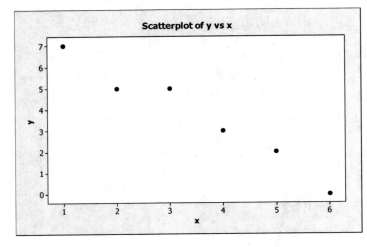

b Calculate

$$S_{xy} = \sum x_i y_i - \frac{(\sum x_i)(\sum y_i)}{n} = 54 - \frac{21(22)}{6} = -23$$

$$S_{xx} = \sum x_i^2 - \frac{(\sum x_i)^2}{n} = 91 - \frac{21^2}{6} = 17.5$$

$$S_{yy} = \sum y_i^2 - \frac{(\sum y_i)^2}{n} = 112 - \frac{22^2}{6} = 31.33333$$

Then $r = \frac{S_{xy}}{\sqrt{S_{xx} S_{yy}}} = \frac{-23}{\sqrt{17.5(31.3333)}} = -0.982$.

c We first calculate the coefficient of determination:

$$r^2 = (-0.982)^2 = 0.9647.$$

This value implies that the sum of squares of deviations is reduced by 96.47% using the linear model $\hat{y} = a + bx$ instead of $\bar{y}$ to predict values of y.

12.53 **a** When x is large, y should be small if the barnacles compete for space on the lobster's surface. Hence, we would expect to find negative correlation.

b-c The test of hypothesis is

$$H_0 : \rho = 0 \quad \text{versus} \quad H_a : \rho < 0$$

Calculate

$$S_{xy} = \sum x_i y_i - \frac{(\sum x_i)(\sum y_i)}{n} = 42,556 - \frac{2379(652)}{10} = -112,554.8$$

$$S_{xx} = \sum x_i^2 - \frac{(\sum x_i)^2}{n} = 973,255 - \frac{2379^2}{10} = 407,290.9$$

$$S_{yy} = \sum y_i^2 - \frac{(\sum y_i)^2}{n} = 114,624 - \frac{652^2}{10} = 102,113.4$$

Then $r = \frac{S_{xy}}{\sqrt{S_{xx} S_{yy}}} = \frac{-112,544.8}{\sqrt{407,290.9(102,113.6)}} = -0.5519$ and the test statistic is

$$t = \frac{r\sqrt{n-2}}{\sqrt{1-r^2}} = \frac{-0.5519\sqrt{8}}{\sqrt{1-(-0.5519)^2}} = -1.872$$

The rejection region for $\alpha = 0.05$ is $t < -t_{.05} = -1.860$ and H_0 is rejected. There is evidence of negative correlation.

12.55 **a** The hypothesis of interest is

$$H_0 : \rho = 0 \quad \text{versus} \quad H_a : \rho \neq 0$$

and the test statistic is

$$t = \frac{r\sqrt{n-2}}{\sqrt{1-r^2}} = \frac{-0.37\sqrt{67}}{\sqrt{1-(-0.37)^2}} = -3.260$$

The rejection region is $|t| > t_{.025} \approx 1.96$ and H_0 is rejected. There is evidence of correlation between x and y.

b The p-value can be bounded using Table 4 as

$$p\text{-value} = 2P(t > 3.26) < 2(0.005) = 0.01$$

c The negative correlation observed above implies that, if the skater's stride is large, his time to completion will be small.

165

12.57   Using a computer program, your scientific calculator or the computing formulas given in the text to calculate the correlation coefficient r.

$$S_{xy} = \sum x_i y_i - \frac{(\sum x_i)(\sum y_i)}{n} = 1,901,500 - \frac{8050(2100)}{9} = 23,166.667$$

$$S_{xx} = \sum x_i^2 - \frac{(\sum x_i)^2}{n} = 7,802,500 - \frac{8050^2}{9} = 602,222.22$$

$$S_{yy} = \sum y_i^2 - \frac{(\sum y_i)^2}{n} = 498,200 - \frac{2100^2}{9} = 8200$$

Then $r = \dfrac{S_{xy}}{\sqrt{S_{xx}S_{yy}}} = \dfrac{23,166.667}{\sqrt{602,222.22(8200)}} = 0.3297$.   The test of hypothesis is

$$H_0 : \rho = 0 \quad \text{versus} \quad H_a : \rho > 0$$

and the test statistic is

$$t = \frac{r\sqrt{n-2}}{\sqrt{1-r^2}} = \frac{0.3297\sqrt{7}}{\sqrt{1-(0.3297)^2}} = 0.92$$

with p-value bounded as

$$p\text{-value} = P(t > 0.92) > 0.10$$

The results are not significant; H_0 is not rejected. There is insufficient evidence to indicate a positive correlation between average maximum drill hole depth and average maximum temperature.

12.59   **a**   Using a computer program, your scientific calculator or the computing formulas given in the text to calculate the correlation coefficient r.

$$S_{xy} = \sum x_i y_i - \frac{(\sum x_i)(\sum y_i)}{n} = 88,140.6 - \frac{1180.3(896)}{12} = 11.5333$$

$$S_{xx} = \sum x_i^2 - \frac{(\sum x_i)^2}{n} = 116,103.03 - \frac{1180.3^2}{12} = 10.689167$$

$$S_{yy} = \sum y_i^2 - \frac{(\sum y_i)^2}{n} = 67,312 - \frac{896^2}{12} = 410.6667$$

Then

$$r = \frac{S_{xy}}{\sqrt{S_{xx}S_{yy}}} = \frac{11.5333}{\sqrt{10.689167(410.6667)}} = 0.1741$$

b   The test of hypothesis is $H_0 : \rho = 0 \quad \text{versus} \quad H_a : \rho \ne 0$
and the test statistic is

$$t = \frac{r\sqrt{n-2}}{\sqrt{1-r^2}} = \frac{0.1741\sqrt{10}}{\sqrt{1-(0.1741)^2}} = 0.559$$

The rejection region with $\alpha = .05$ is $|t| > t_{.025} = 2.228$ and H_0 is not rejected. There is insufficient evidence to indicate a correlation between body temperature and heart rate.

12.61   **a**   The calculations shown below are done using the computing formulas. An appropriate computer program will provide identical results to within rounding error.

$$\sum x_i = 720 \qquad \sum y_i = 324 \qquad \sum x_i y_i = 17,540$$
$$\sum x_i^2 = 49,200 \qquad \sum y_i^2 = 9540 \qquad n = 12$$

Then

$$S_{xy} = \sum x_i y_i - \frac{(\sum x_i)(\sum y_i)}{n} = 17,540 - \frac{720(324)}{12} = -1900$$

$$S_{xx} = \sum x_i^2 - \frac{(\sum x_i)^2}{n} = 49,200 - \frac{720^2}{12} = 6000$$

$$S_{yy} = \sum y_i^2 - \frac{(\sum y_i)^2}{n} = 9540 - \frac{324^2}{12} = 792$$

$$b = \frac{S_{xy}}{S_{xx}} = \frac{-1900}{6000} = -0.317 \text{ and } a = \bar{y} - b\bar{x} = 27 - (-0.317)(60) = 46.000$$

and the least squares line is

$$\hat{y} = a + bx = 46 - 0.317x.$$

    b   The graph of the least squares line and the 12 data points are shown below.

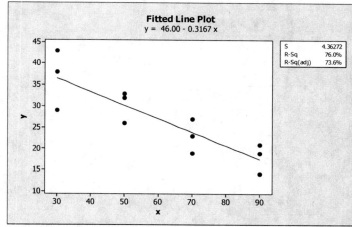

    c   Since Total SS $= S_{yy} = 792$ and

$$SSR = \frac{\left(S_{xy}\right)^2}{S_{xx}} = \frac{(-1900)^2}{6000} = 601.6667$$

Then

$$SSE = \text{Total SS} - SSR = S_{yy} - \frac{\left(S_{xy}\right)^2}{S_{xx}} = 190.3333$$

The ANOVA table with 1 *df* for regression and $n - 2$ *df* for error is shown below. Remember that the mean squares are calculated as $MS = SS/df$.

Source	*df*	SS	MS
Regression	1	601.6667	601.6667
Error	10	190.3333	19.0333
Total	11	792.0000	

    d   The diagnostic plots for the regression analysis are shown below. Although the second plot indicates that the responses are slightly more variable when $x = 90$, and the normality plot is slightly irregular, these irregularities are probably not significant, given the small sample size.

167

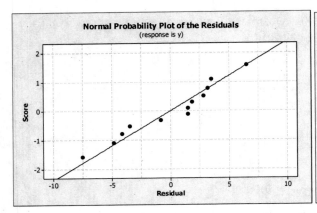 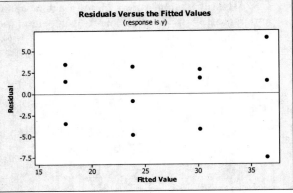

e   The 95% confidence interval for β is

$$b \pm t_{.025}\sqrt{\frac{MSE}{S_{xx}}} \Rightarrow -0.317 \pm 2.228\sqrt{\frac{19.0333}{6000}} \Rightarrow -0.317 \pm 0.125$$

or $-0.442 < \beta < -0.192$.

f   When $x = 50$, the estimate of mean potency $E(y)$ is $\hat{y} = 46.00 - 0.317(50) = 30.167$ and the 95% confidence interval is

$$\hat{y} \pm t_{.025}\sqrt{MSE\left(\frac{1}{n} + \frac{(x_p - \bar{x})^2}{S_{xx}}\right)}$$

$$30.167 \pm 2.228\sqrt{19.0333\left(\frac{1}{12} + \frac{(50-60)^2}{6000}\right)}$$

$$30.167 \pm 3.074$$

or $27.09 < E(y) < 33.24$.

g   The predictor for y when $x = 50$ is $\hat{y} = 30.167$ and the 95% prediction interval is

$$\hat{y} \pm t_{.025}\sqrt{MSE\left(1 + \frac{1}{n} + \frac{(x_p - \bar{x})^2}{S_{xx}}\right)}$$

$$30.167 \pm 2.228\sqrt{19.0333\left(1 + \frac{1}{12} + \frac{(50-60)^2}{6000}\right)}$$

$$30.167 \pm 10.195$$

or $19.97 < y < 40.36$.

12.63   Answers will vary. The *Minitab* output for this linear regression problem is shown on the next page.

Regression Analysis: y versus x

```
The regression equation is
y = 0.067 + 0.517 x

Predictor      Coef    SE Coef     T      P
Constant     0.0667    0.3935    0.17   0.870
x            0.51667   0.09107   5.67   0.001
S = 0.446148   R-Sq = 82.1%   R-Sq(adj) = 79.6%

Analysis of Variance
Source          DF      SS       MS       F      P
Regression       1    6.4067   6.4067   32.19  0.001
Residual Error   7    1.3933   0.1990
Total            8    7.8000
```

The printout indicates a significant linear regression ($t = 5.67$) with the regression line given as $\hat{y} = 0.067 + 0.517x$.

12.65   **a**   Notice that all three regressions have p-values that are very small. That is each of the three regression analyses is significant.

b   Based on the values of the coefficients of determination, it appears that radiographs ($r^2 = 0.80$) are the most effective, followed by 3-D MRIs ($r^2 = 0.65$) and standard MRIs ($r^2 = 0.43$).

c   The p-values provide consistent conclusions, except that we cannot differentiate between radiographs and 3-D MRIs (the p-values are both < 0.0001.)

12.67   **a-b**   Answers will vary. The *Minitab* printout is shown below, along with two diagnostic plots. These plots show one unusual observation, giving it an unusually large influence on the estimate of the regression line. The printout indicates a significant linear regression ($t = 36.21$ with p-value $= .000$) and the regression line is given as $\hat{y} = 0.546 + 0.974x$.

Regression Analysis: Actual versus Estimated

```
The regression equation is
Actual = 0.546 + 0.974 Estimated

Predictor      Coef    SE Coef     T      P
Constant     0.5457    0.4097    1.33   0.220
Estimated    0.97410   0.02690  36.21   0.000
S = 0.942569   R-Sq = 99.4%   R-Sq(adj) = 99.3%

Analysis of Variance
Source          DF      SS       MS       F       P
Regression       1    1164.8   1164.8  1311.09  0.000
Residual Error   8       7.1      0.9
Total            9    1171.9

Unusual Observations
Obs   Estimated   Actual     Fit   SE Fit   Residual   St Resid
  7        42.0   41.500   41.458   0.900      0.042      0.15 X

X denotes an observation whose X value gives it large influence.
```

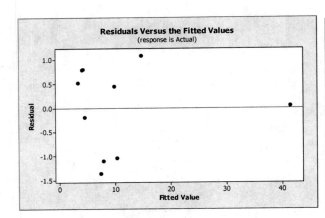

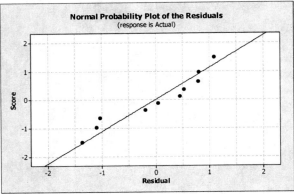

12.69 The least squares line is given in the printout as $\hat{y} = 18.427 + 8.1768x$, and the coefficient of determination is $r^2 = 0.939$. That is, 93.9% of the total variation in y can be explained by the independent variable x. The linear model fits the data fairly well.

12.71 Use a computer, your scientific calculator or the computing formulas to find the correlation between x and y.

$$S_{xy} = \sum x_i y_i - \frac{(\sum x_i)(\sum y_i)}{n} = 764.74 - \frac{92.8(86.4)}{11} = 35.83818$$

$$S_{xx} = \sum x_i^2 - \frac{(\sum x_i)^2}{n} = 852.12 - \frac{92.8^2}{11} = 69.22545$$

$$S_{yy} = \sum y_i^2 - \frac{(\sum y_i)^2}{n} = 737.42 - \frac{86.4^2}{11} = 58.78727$$

Then $r = \dfrac{S_{xy}}{\sqrt{S_{xx}S_{yy}}} = \dfrac{35.83818}{\sqrt{69.22545(58.78727)}} = 0.5618$.

The test of hypothesis is

$$H_0 : \rho = 0 \quad \text{versus} \quad H_a : \rho > 0$$

and the test statistic is

$$t = \frac{r\sqrt{n-2}}{\sqrt{1-r^2}} = \frac{0.5618\sqrt{9}}{\sqrt{1-(0.5618)^2}} = 2.04$$

with p-value $= P(t > 2.04)$ bounded as

$$0.025 < p\text{-value} < 0.05$$

Hence, at the 5% level of significance, H_0 is rejected. There is sufficient evidence to indicate a positive correlation between the changes in bodyweight for the pairs of twins. The *Minitab* correlation printout below shows $r = 0.562$ with *two-tailed p*-value $= .072$ so that the *one-tailed p*-value would be $.072/2 = .036$ which is significant at the 5% level of significance. Again, you can conclude that there is a significant positive correlation between the changes in body weight for the pairs of twins.

Correlations: x, y
```
Pearson correlation of x and y = 0.562
P-Value = 0.072
```

12.73 The fitted line may not adequately describe the relationship between x and y outside the experimental region.

12.75 The calculations shown below are done using the computing formulas. An appropriate computer program will provide identical results to within rounding error.

$$\sum x_i = 20 \qquad \sum y_i = 364 \qquad \sum x_i y_i = 1064$$
$$\sum x_i^2 = 60 \qquad \sum y_i^2 = 18,984 \qquad n = 8$$

Then

$$S_{xy} = \sum x_i y_i - \frac{\left(\sum x_i\right)\left(\sum y_i\right)}{n} = 1064 - \frac{20(364)}{8} = 154$$

$$S_{xx} = \sum x_i^2 - \frac{\left(\sum x_i\right)^2}{n} = 60 - \frac{20^2}{8} = 10$$

$$S_{yy} = \sum y_i^2 - \frac{\left(\sum y_i\right)^2}{n} = 18,984 - \frac{364^2}{8} = 2422$$

a $\quad b = \dfrac{S_{xy}}{S_{xx}} = \dfrac{154}{10} = 15.4$ and $a = \bar{y} - b\bar{x} = 45.5 - (15.4)(2.5) = 7$

and the least squares line is

$$\hat{y} = a + bx = 7 + 15.4x.$$

b $\quad$ Since Total $SS = S_{yy} = 2422$ and

$$SSR = \frac{\left(S_{xy}\right)^2}{S_{xx}} = \frac{(154)^2}{10} = 2371.6$$

Then $\qquad SSE = \text{Total } SS - SSR = S_{yy} - \dfrac{\left(S_{xy}\right)^2}{S_{xx}} = 50.4$

The ANOVA table with 1 df for regression and $n - 2$ df for error is shown below. Remember that the mean squares are calculated as $MS = SS/df$.

Source	df	SS	MS
Regression	1	2371.6	2371.6
Error	6	50.4	8.4
Total	7	2422.0	

c $\quad$ To test $H_0 : \beta = 0, H_a : \beta \neq 0$, the test statistic is

$$t = \frac{b - \beta_0}{s/\sqrt{S_{xx}}} = \frac{15.4}{\sqrt{8.4/10}} = 16.80$$

The rejection region for $\alpha = 0.05$ is $|t| > t_{.025} = 2.447$ and we reject H_0. There is sufficient evidence to indicate that yield and amount of nitrogen are linearly related.

d $\quad$ When $x = 2.5$, the estimate of average yield $E(y)$ is $\hat{y} = 7 + 15.4(2.5) = 45.5$ and the 95% confidence interval is

$$\hat{y} \pm t_{.025} \sqrt{MSE\left(\frac{1}{n} + \frac{\left(x_p - \bar{x}\right)^2}{S_{xx}}\right)}$$

$$45.5 \pm 2.447 \sqrt{8.4\left(\frac{1}{8} + \frac{(2.5 - 2.5)^2}{10}\right)}$$

$$45.5 \pm 2.51$$

or $\quad 42.99 < E(y) < 48.01$.

e $\quad$ The average increase in y for a one-unit (100 pound) increase in x is the slope of the line. The 95% confidence interval for β is

$$b \pm t_{.025} \sqrt{\frac{MSE}{S_{xx}}} \Rightarrow 15.4 \pm 2.447 \sqrt{\frac{8.4}{10}} \Rightarrow 15.4 \pm 2.24$$

or $13.16 < \beta < 17.64$.

f Calculate
$$r^2 = \frac{SSR}{\text{Total SS}} = \frac{2371.6}{2422} = 0.979$$
The total variation has been reduced by 97.9%% by using the linear model.

12.77 **a** The scatterplot is shown below. There is a strong *curvilinear* relationship between sales and year.

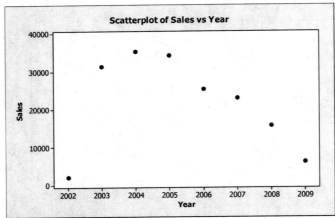

b Even though the scatterplot indicates that a linear relationship may not be correct, we fit the least-squares regression line anyway. The calculations shown below are done using the computing formulas. An appropriate computer program will provide identical results to within rounding error.

$$\sum x_i = 16,044 \qquad \sum y_i = 173,808 \qquad \sum x_i y_i = 348,524,039$$
$$\sum x_i^2 = 32,176,284 \qquad \sum y_i^2 = 4,888,731,444 \qquad n = 8$$

Then
$$S_{xy} = \sum x_i y_i - \frac{(\sum x_i)(\sum y_i)}{n} = -47,905$$

$$S_{xx} = \sum x_i^2 - \frac{(\sum x_i)^2}{n} = 42$$

$$S_{yy} = \sum y_i^2 - \frac{(\sum y_i)^2}{n} = 1,112,578,836$$

$$b = \frac{S_{xy}}{S_{xx}} = \frac{-47,905}{42} = -1140.595238 \text{ and } a = \overline{y} - b\overline{x} = 21,726 - (-1140.595238)(2005.5) = 2,309,189.75$$

and the least squares line is
$$\hat{y} = 2,309,189.75 - 1140.595x.$$

c Since Total SS $= S_{yy} = 1,112,578,836$ and

$$SSR = \frac{(S_{xy})^2}{S_{xx}} = \frac{(-47,905)^2}{42} = 54,640,214.88$$

Then
$$SSE = \text{Total SS} - SSR = S_{yy} - \frac{(S_{xy})^2}{S_{xx}} = 1,057,938,621.12$$

and MSE $= 1,057,938,621.12/6 = 176,323,103.5$. To test $H_0 : \beta = 0, H_a : \beta \neq 0$, the test statistic is
$$t = \frac{b - \beta_0}{s/\sqrt{S_{xx}}} = \frac{-1140.595238}{\sqrt{MSE/42}} = -.56$$

172

The rejection region for $\alpha = 0.05$ is $|t| > t_{.025} = 2.447$ and we do not reject H$_0$. There is insufficient evidence to indicate that sales and year are linearly related.

d-e   The diagnostic plots are given in the text. The residual versus fits plot suggests that there may be a quadratic effect, and that the linear model fit in part **b** is incorrect. Therefore, it is not advisable to predict the 2010 sales using the regression line from part **b.**

13: Multiple Regression Analysis

13.1 **a** When $x_2 = 2$, $E(y) = 3 + x_1 - 2(2) = x_1 - 1$.

When $x_2 = 1$, $E(y) = 3 + x_1 - 2(1) = x_1 + 1$.

When $x_2 = 0$, $E(y) = 3 + x_1 - 2(0) = x_1 + 3$.

These three straight lines are graphed below.

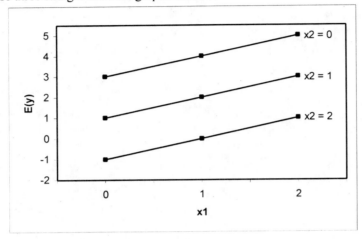

b Notice that the lines are parallel (they have the same slope).

13.3 **a** The hypothesis to be tested is

$$H_0 : \beta_1 = \beta_2 = \beta_3 = 0 \qquad H_a\text{: at least one } \beta_i \text{ differs from zero}$$

and the test statistic is

$$F = \frac{\text{MSR}}{\text{MSE}} = 57.44$$

which has an F distribution with $df_1 = k = 3$ and $df_2 = n - k - 1 = 15 - 3 - 1 = 11$. The rejection region for $\alpha = .05$, which is found in the upper tail of the F distribution, is $F > 3.59$ and H_0 is rejected. Alternatively, the p-value with $df_1 = k = 3$ and $df_2 = n - k - 1 = 15 - 3 - 1 = 11$ can be bounded (using Table 6 in Appendix I) as p-value $< .005$. There is evidence that the model contributes information for the prediction of y.

b Use the fact that

$$F = \frac{R^2/k}{(1 - R^2)/[n - (k + 1)]} = 57.44$$

Solving for R^2 you find

$$\frac{R^2/3}{(1 - R^2)/11} = 57.44$$

$$.33R^2 = 5.2218 - 5.2218R^2$$

$$R^2 = \frac{5.2218}{5.5551} = .94$$

If $R^2 = .94$, the total sum of squares of deviations of the y-values about their mean has been reduced by 94% by using the linear model to predict y.

174

13.5　**a**　The model is quadratic.

b　Since $R^2 = .815$, the sum of squares of deviations is reduced by 81.5% using the quadratic model rather than $\bar{y}$ to predict y.

c　The hypothesis to be tested is

$$H_0 : \beta_1 = \beta_2 = 0 \qquad H_a: \text{at least one } \beta_i \text{ differs from zero}$$

and the test statistic is

$$F = \frac{\text{MSR}}{\text{MSE}} = 37.37$$

which has an F distribution with $df_1 = k = 2$ and $df_2 = n - k - 1 = 20 - 2 - 1 = 17$. The p-value given in the printout is $P = .000$ and H_0 is rejected. There is evidence that the model contributes information for the prediction of y.

13.7　**a**　Refer to the printout in Exercise 13.5. When $x = 0$, the estimate of $E(y)$ is

$$\hat{y} = 10.5638 + 4.4366(0) - .64754(0)^2 = 10.5638.$$

b　Since $E(y) = \beta_0 + \beta_1 x + \beta_2 x^2$, when $x = 0$, $E(y) = \beta_0$. A test of $E(y \text{ given } x = 0) = 0$ is equivalent to a test of

$$H_0 : \beta_0 = 0 \qquad H_a: \beta_0 \neq 0$$

The individual t-test is

$$t = \frac{b_0}{SE(b_0)} = \frac{10.5638}{.6951} = 15.20$$

with p-value $= .000$ and H_0 is rejected. The mean value of y differs from zero when $x = 0$.

13.9　**a**　Rate of increase is measured by the slope of a line tangent to the curve; this line is given by an equation obtained as dy/dx, the derivative of y with respect to x. In particular,

$$\frac{dy}{dx} = \frac{d}{dx}\left(\beta_0 + \beta_1 x + \beta_2 x^2\right) = \beta_1 + 2\beta_2 x$$

which has slope $2\beta_2$. If β_2 is negative, then the rate of increase is decreasing. Hence, the hypothesis of interest is

$$H_0 : \beta_2 = 0, \qquad H_a: \beta_2 < 0$$

b　The individual t-test is $t = -8.11$ as in Exercise 13.8b. However, the test is one-tailed, which means that the p-value is half of the amount given in the printout. That is, $p\text{-value} = \frac{1}{2}(.000) = .000$. Hence, H_0 is again rejected. There is evidence to indicate a decreasing rate of increase.

13.11　Refer to the *Excel* printout in Exercise 13.10.

a　From the printout, $SSR = 234.955$ and $\text{Total SS} = S_{yy} = 236.015$. Then

$$R^2 = \frac{SSR}{\text{Total SS}} = \frac{234.955}{236.015} = .9955$$

which agrees with the printout.

b　Calculate $R^2(\text{adj}) = \left(1 - \frac{\text{MSE}}{\text{Total SS}/(n-1)}\right)100\% = \left(1 - \frac{.353}{236.015/5}\right)100\% = 99.25\%$

The value of R^2(adj) can be used to compare two or more regression models using different numbers of independent predictor variables. Since the value of $R^2(\text{adj}) = 99.25\%$ is just slightly larger than the value of $R^2(\text{adj}) = 95.66\%$ for the linear model, the quadratic model fits just slightly better.

13.13 **a** The values of R^2(adj) should be used to compare several different regression models. For the eight possible models given in the *Minitab* output, the largest value of R^2(adj) is 39.5% which occurs when the following model is fit to the data:

$$y = \beta_0 + \beta_1 x_1 + \beta_2 x_3 + \beta_3 x_5 + \varepsilon$$

b Since the values of R^2 and R^2(adj) are both very small, even the best of the models is not very useful for predicting taste score based on these three independent variables.

13.15 **a** The *Minitab* printout fitting the model to the data is shown below. The least squares line is

$$\hat{y} = -8.177 + 0.292 x_1 + 4.434 x_2$$

Regression Analysis: y versus x1, x2

```
The regression equation is
y = - 8.18 + 0.292 x1 + 4.43 x2

Predictor     Coef   SE Coef       T       P
Constant    -8.177     4.206   -1.94   0.093
x1          0.2921    0.1357    2.15   0.068
x2          4.4343    0.8002    5.54   0.001
S = 3.30335   R-Sq = 82.3%   R-Sq(adj) = 77.2%

Analysis of Variance
Source           DF       SS      MS       F       P
Regression        2   355.22  177.61   16.28   0.002
Residual Error    7    76.38   10.91
Total             9   431.60

Source  DF   Seq SS
x1       1    20.16
x2       1   335.05
```

b The F test for the overall utility of the model is $F = 16.28$ with $P = .002$. The results are highly significant; the model contributes significant information for the prediction of y.

c To test the effect of advertising expenditure, the hypothesis of interest is

$$H_0 : \beta_2 = 0, \qquad H_a : \beta_2 \neq 0$$

and the test statistic is $t = 5.54$ with p-value $= .001$. Since $\alpha = .01$, H_0 is rejected. We conclude that advertising expenditure contributes significant information for the prediction of y, given that capital investment is already in the model.

d From the *Minitab* printout, R-Sq = 82.3%, which means that 82.3% of the total variation can be explained by the quadratic model. The model is very effective.

13.17 **a** Quantitative   **b** Quantitative
       c Qualitative ($x_1 = 1$ if B; 0 otherwise $x_2 = 1$ if C; 0 otherwise)
       d Quantitative   **e** Qualitative ($x_1 = 1$ if day shift; 0 otherwise)

13.19 **a** The variable x_2 must be the quantitative variable, since it appears as a quadratic term in the model. Qualitative variables appear only with exponent 1, although they may appear as the coefficient of another quantitative variable with exponent 2 or greater.

b When $x_1 = 0$, $\hat{y} = 12.6 + 3.9 x_2^2$ while when $x_1 = 1$,

$$\hat{y} = 12.6 + .54(1) - 1.2 x_2 + 3.9 x_2^2$$
$$= 13.14 - 1.2 x_2 + 3.9 x_2^2$$

c The graph on the next page shows the two parabolas.

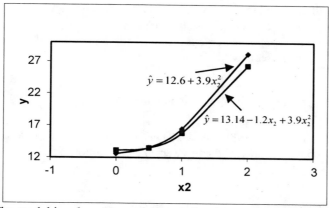

13.21   **a**   The model involves two variables and an interaction between the two:

$$E(y) = \beta_0 + \beta_1 x_1 + \beta_2 x_2 + \beta_3 x_1 x_2$$

where $x_1 = 0$ if cotton; 1 if cucumber and x_2 = temperature.

b   The *Minitab* regression printout is shown below.

Regression Analysis: y versus x1, x2, x1x2

```
The regression equation is
y = - 9.4 - 29.4 x1 + 0.630 x2 + 0.540 x1x2

Predictor     Coef    SE Coef      T      P
Constant     -9.39      47.14  -0.20  0.844
x1          -29.42      64.73  -0.45  0.654
x2          0.6302     0.6244   1.01  0.324
x1x2        0.5405     0.8543   0.63  0.533
S = 10.5727   R-Sq = 37.8%   R-Sq(adj) = 29.3%

Analysis of Variance
Source          DF      SS      MS      F      P
Regression       3  1493.3   497.8   4.45  0.014
Residual Error  22  2459.2   111.8
Total           25  3952.5

Source  DF   Seq SS
x1       1    928.6
x2       1    519.9
x1x2     1     44.7
```

c   Look first at the interaction effect. The interaction term is not significant ($t = .63$ with $P = .533$). That is, there is insufficient evidence to indicate that the effect of temperature on the number of eggs is different depending on the type of plant.

d   Since the interaction term is not significant, it could be removed and the data refit using the model

$$E(y) = \beta_0 + \beta_1 x_1 + \beta_2 x_2$$

The *Minitab* printout for the regression analysis with interaction removed is shown below.

Regression Analysis: y versus x1, x2

```
The regression equation is
y = - 31.1 + 11.4 x1 + 0.919 x2

Predictor     Coef    SE Coef      T      P
Constant    -31.15      31.82  -0.98  0.338
x1          11.441      4.112   2.78  0.011
x2          0.9190     0.4205   2.19  0.039
S = 10.4339   R-Sq = 36.6%   R-Sq(adj) = 31.1%

Analysis of Variance
Source          DF      SS      MS      F      P
Regression       2  1448.5   724.3   6.65  0.005
Residual Error  23  2503.9   108.9
Total           25  3952.5
```

Notice that both variables are significant, and that the overall model contributes significant information for the prediction of y. However, since $R^2 = 36.6\%$, there is still much variation which has not been accounted for. The model without interaction is better, but still does not fit as well as it might. Perhaps there are other variables that the experimenter should explore.

e Answers will vary.

13.23 The basic response equation for a specific type of bonding compound would be
$$E(y) = \beta_0 + \beta_1 x_1 + \beta_2 x_1^2$$
Since the qualitative variable "bonding compound" is at two levels, one dummy variable is needed to incorporate this variable into the model. Define the dummy variable x_2 as follows:
$$x_2 = 1 \text{ if bonding compound 2}$$
$$= 0 \text{ otherwise}$$
The expanded model is now written as
$$E(y) = \beta_0 + \beta_1 x_1 + \beta_2 x_1^2 + \beta_3 x_2 + \beta_4 x_1 x_2 + \beta_5 x_1^2 x_2 \text{ or}$$
$$y = \beta_0 + \beta_1 x_1 + \beta_2 x_1^2 + \beta_3 x_2 + \beta_4 x_1 x_2 + \beta_5 x_1^2 x_2 + \varepsilon$$

13.25 **a** From the printout, the prediction equation is $\hat{y} = 8.585 + 3.8208x - 0.21663x^2$.

b R^2 is labeled "R-sq" or $R^2 = .944$. Hence 94.4% of the total variation is accounted for by using x and x^2 in the model.

c The hypothesis of interest is
$$H_0 : \beta_1 = \beta_2 = 0 \quad H_a : \text{at least one } \beta_i \text{ differs from zero}$$
and the test statistic is $F = 33.44$ with p-value $= .003$. Hence, H_0 is rejected, and we conclude that the model contributes significant information for the prediction of y.

d The hypothesis of interest is
$$H_0 : \beta_2 = 0 \quad H_a : \beta_2 \neq 0$$
and the test statistic is $t = -4.93$ with p-value $= .008$. Hence, H_0 is rejected, and we conclude that the quadratic model provides a better fit to the data than a simple linear model.

e The pattern of the diagnostic plots does not indicate any obvious violation of the regression assumptions.

13.27 The *Minitab* printout for the data is shown below.

Regression Analysis: y versus x1, x2, x3, x1x2, x1x3

```
The regression equation is
y = 4.10 + 1.04 x1 + 3.53 x2 + 4.76 x3 - 0.430 x1x2 - 0.080 x1x3

Predictor      Coef   SE Coef       T       P
Constant     4.1000    0.3860   10.62   0.000
x1           1.0400    0.1164    8.94   0.000
x2           3.5300    0.5459    6.47   0.000
x3           4.7600    0.5459    8.72   0.000
x1x2        -0.4300    0.1646   -2.61   0.028
x1x3        -0.0800    0.1646   -0.49   0.639
S = 0.368028   R-Sq = 98.4%   R-Sq(adj) = 97.5%

Analysis of Variance
Source          DF       SS       MS       F       P
Regression       5   74.830   14.966  110.50   0.000
Residual Error   9    1.219    0.135
Total           14   76.049
```

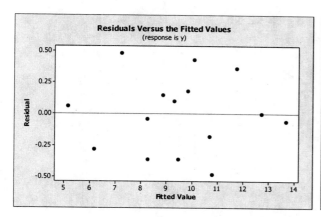

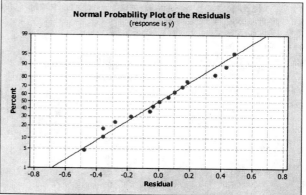

a The model fits very well, with an overall $F = 110.50$ (P = .000) and $R^2 = .984$. The diagnostic plots indicate no violations of the regression assumptions.

b The parameter estimates are found in the column marked "Coef" and the prediction equation is

$$\hat{y} = 4.10 + 1.04x_1 + 3.53x_2 + 4.76x_3 - 0.43x_1x_2 - 0.08x_1x_3$$

Using the dummy variables defined in Exercise 13.26, the coefficients can be combined to give the three lines that are graphed in the figure below.

Men: $\hat{y} = 4.10 + 1.04x_1$

Children: $\hat{y} = 7.63 + 0.61x_1$

Women: $\hat{y} = 8.86 + 0.96x_1$

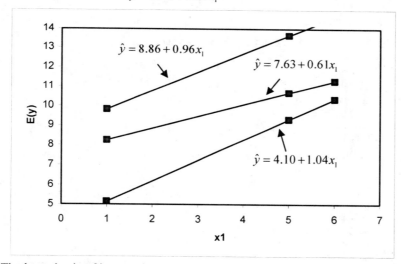

c The hypothesis of interest is

$$H_0 : \beta_4 = 0 \qquad H_a : \beta_4 \neq 0$$

and the test statistic is $t = -2.61$ with P = .028. Since this value is less than .05, the results are significant at the 5% level of significance and H_0 is rejected. There is a difference in the slopes.

d The hypothesis of interest is

$$H_0 : \beta_4 = \beta_5 = 0 \qquad H_a : \text{at least one } \beta_i \text{ differs from zero for } i = 4, 5$$

Using the methods of Section 13.5 and the *Minitab* printout above, $SSE_2 = 1.219$ with 9 degrees of freedom, while the printout on the next page, fit using the reduced model gives $SSE_1 = 2.265$ with 11 degrees of freedom.

Regression Analysis (reduced model): y versus x1, x2, x3

```
The regression equation is
y = 4.61 + 0.870 x1 + 2.24 x2 + 4.52 x3

Predictor       Coef   SE Coef       T       P
Constant      4.6100    0.3209   14.37   0.000
x1            0.87000   0.08285  10.50   0.000
x2            2.2400    0.2870    7.81   0.000
x3            4.5200    0.2870   15.75   0.000

S = 0.453772   R-Sq = 97.0%   R-Sq(adj) = 96.2%

Analysis of Variance
Source          DF       SS      MS       F       P
Regression       3   73.784  24.595  119.44   0.000
Residual Error  11    2.265   0.206
Total           14   76.049
```

Hence, the degrees of freedom associated with $SSE_1 - SSE_2 = 1.046$ is $11 - 9 = 2$. The test statistic is

$$F = \frac{(SSE_1 - SSE_2)/2}{SSE_2/9} = \frac{1.046/2}{.1354} = 3.86$$

The rejection region with $\alpha = .05$ is $F > F_{.05} = 4.26$ (with 2 and 9 df) and H_0 is not rejected. The interaction terms in the model are not significant. The experimenter should consider eliminating these terms from the model.

e Answers will vary.

13.29 **a** The model is

$$y = \beta_0 + \beta_1 x_1 + \beta_2 x_2 + \beta_3 x_1^2 + \beta_4 x_1 x_2 + \beta_5 x_1^2 x_2 + \varepsilon$$

and the *Minitab* printout is shown below.

Regression Analysis: y versus x1, x2, x1sq, x1x2, x1sqx2

```
The regression equation is
y = 4.5 + 6.39 x1 - 50.9 x2 + 0.132 x1sq + 17.1 x1x2 - 0.502 x1sqx2

Predictor       Coef   SE Coef       T       P
Constant       4.51     42.24    0.11   0.916
x1             6.394     5.777    1.11   0.275
x2           -50.85     56.21   -0.90   0.371
x1sq           0.1318    0.1687   0.78   0.439
x1x2          17.064     7.101    2.40   0.021
x1sqx2        -0.5025    0.1992  -2.52   0.016
S = 71.6891   R-Sq = 76.8%   R-Sq(adj) = 73.8%

Analysis of Variance
Source          DF       SS       MS       F       P
Regression       5   664164   132833   25.85   0.000
Residual Error  39   200434     5139
Total           44   864598
```

b The fitted prediction model uses the coefficients given in the column marked "Coef" in the printout:

$$\hat{y} = 4.51 + 6.394x_1 - 50.85x_2 + 17.064x_1 x_2 + .1318x_1^2 - .5025x_1^2 x_2$$

The F test for the model's utility is $F = 25.85$ with P $= .000$ and $R^2 = .768$. The model fits quite well.

c If the dolphin is female, $x_2 = 0$ and the prediction equation becomes

$$\hat{y} = 4.51 + 6.394x_1 + .1318x_1^2$$

d If the dolphin is male, $x_2 = 1$ and the prediction equation becomes

$$\hat{y} = -46.34 + 23.458x_1 - .3707x_1^2$$

e The hypothesis of interest is

$$H_0 : \beta_4 = 0 \qquad H_a : \beta_4 \neq 0$$

and the test statistic is $t = .78$ with p-value $= .439$. H_0 is not rejected and we conclude that the quadratic term is not important in predicting mercury concentration for female dolphins.

180

13.31 **a** The correlation matrix for Cost, y, x_1, x_2, x_3, x_4, and x_5 was obtained using *Minitab* and is shown below:

Correlations: Price, y, x1, x2, x3, x4, x5

	Price	y	x1	x2	x3	x4
y	0.609					
	0.001					
x1	0.385	0.614				
	0.052	0.001				
x2	0.059	0.129	0.044			
	0.774	0.530	0.831			
x3	0.170	0.396	0.161	0.297		
	0.407	0.045	0.433	0.140		
x4	-0.036	0.091	0.183	-0.333	-0.177	
	0.860	0.660	0.370	0.096	0.388	
x5	0.136	0.297	0.321	0.000	0.333	-0.177
	0.508	0.141	0.110	1.000	0.096	0.388

```
Cell Contents: Pearson correlation
               P-Value
```

Scanning down the first column labeled "Price", you can see that only y has a p-value less than 0.05, so that only y = overall score is significantly correlated with price. However, scanning down the second column labeled "y", you will find that x_1 and x_3 both have p-values less than 0.05, indicating a significant correlation with y.

b-c Using all five independent variables, the model is
$$E(y) = \beta_0 + \beta_1 x_1 + \beta_2 x_2 + \beta_3 x_3 + \beta_4 x_4 + \beta_5 x_5$$
and the *Minitab* printout is shown below.

Regression Analysis: y versus x1, x2, x3, x4, x5
```
The regression equation is
y = 33.1 + 7.04 x1 + 0.45 x2 + 4.00 x3 + 0.45 x4 + 0.23 x5

Predictor    Coef   SE Coef     T      P
Constant    33.10    13.00    2.55   0.019
x1           7.042    2.338   3.01   0.007
x2           0.452    2.190   0.21   0.839
x3           4.002    2.455   1.63   0.119
x4           0.446    1.311   0.34   0.738
x5           0.234    1.262   0.19   0.855
S = 5.37026   R-Sq = 47.1%   R-Sq(adj) = 33.9%
```

Notice that $R^2 = 47.1\%$ and $R^2(adj) = .339\%$. Hence, only 47.1% of the total variation is accounted for by using this regression model, so that the model is only partially explaining the inherent variability in y.

d From the *Minitab* printout in part **b**, you can see that x_1 (p-value = .007) and to some degree x_3 (p-value = .119) appear to be important predictors of y given the other variables in the model. The variables x_2 (p-value = .839), x_4 (p-value = .738) and x_5 (p-value = .855) are not important. If they are removed, the reduced model is
$$E(y) = \beta_0 + \beta_1 x_1 + \beta_2 x_3$$
and the *Minitab* printout for the reduced model is shown below:

Regression Analysis: y versus x1, x3
```
The regression equation is
y = 35.8 + 7.34 x1 + 4.11 x3

Predictor    Coef   SE Coef     T      P
Constant    35.800   8.603    4.16   0.000
x1           7.340    2.003   3.67   0.001
x3           4.110    2.078   1.98   0.060
S = 5.02448   R-Sq = 46.8%   R-Sq(adj) = 42.1%
```

Notice that $R^2(adj) = 42.1\%$, which means that the reduced model is even better than the full model for explaining the inherent variability in y.

13.33 The hypothesis of interest is

$$H_0 : \beta_1 = \beta_2 = \beta_3 = 0$$

and the test statistic is $F = 1.49$ with p-value $= .235$. H_0 is not rejected and we conclude that the model does not provide valuable information for the prediction of y. This matches the results of the analysis of variance F-test.

13.35 **a** $R^2 = .9985$. Hence, 99.85% of the total variation is accounted for by using x and x^2 in the model.
   b The hypothesis of interest is

$$H_0 : \beta_1 = \beta_2 = 0$$

and the test statistic is $F = 1676.610$ with p-value $= .000$. H_0 is rejected and we conclude that the model provides valuable information for the prediction of y.
   c The hypothesis of interest is

$$H_0 : \beta_1 = 0$$

and the test statistic is $t = -2.652$ with p-value $= .045$. H_0 is rejected and we conclude that the linear regression coefficient is significant when x^2 is in the model.
   d The hypothesis of interest is

$$H_0 : \beta_2 = 0$$

and the test statistic is $t = 15.138$ with p-value $= .000$. H_0 is rejected and we conclude that the quadratic regression coefficient is significant when x is in the model.
   e When the quadratic term is removed from the model, the value of $R^2(\text{adj})$ increases by $99.79 - 91.87 = 7.92\%$. The quadratic term makes a significant contribution when it is included in the model.
   f The clear pattern of a curve in the residual plot indicates that the quadratic term should be included in the model.

14: Analysis of Categorical Data

14.1 See Section 14.1 of the text.

14.3 For a test of specified cell probabilities, the degrees of freedom are $k-1$. Use Table 5, Appendix I:
   a $df = 6$; $\chi_{.05}^2 = 12.59$; reject H_0 if $X^2 > 12.59$
   b $df = 9$; $\chi_{.01}^2 = 21.666$; reject H_0 if $X^2 > 21.666$

14.5 **a** Three hundred responses were each classified into one of five categories. The objective is to determine whether or not one category is preferred over another. To see if the five categories are equally likely to occur, the hypothesis of interest is

$$H_0 : p_1 = p_2 = p_3 = p_4 = p_5 = \frac{1}{5}$$

versus the alternative that at least one of the cell probabilities is different from 1/5.

b The number of degrees of freedom is equal to the number of cells, k, less one degree of freedom for each linearly independent restriction placed on $p_1, p_2, \ldots, p_k$. For this exercise, $k = 5$ and one degree of freedom is lost because of the restriction that

$$\sum p_i = 1$$

Hence, X^2 has $k - 1 = 4$ degrees of freedom.

c The rejection region for this test is located in the upper tail of the chi-square distribution with $df = 4$. From Table 5, the appropriate upper-tailed rejection region is $X^2 > \chi_{.05}^2 = 9.4877$.

d The test statistic is

$$X^2 = \sum \frac{(O_i - E_i)^2}{E_i}$$

which, when n is large, possesses an approximate chi-square distribution in repeated sampling. The values of O_i are the actual counts *observed* in the experiment, and

$$E_i = np_i = 300(1/5) = 60.$$

A table of observed and expected cell counts follows:

Category	1	2	3	4	5
O_i	47	63	74	51	65
E_i	60	60	60	60	60

Then

$$X^2 = \frac{(47-60)^2}{60} + \frac{(63-60)^2}{60} + \frac{(74-60)^2}{60} + \frac{(51-60)^2}{60} + \frac{(65-60)^2}{60}$$

$$= \frac{480}{60} = 8.00$$

e Since the observed value of X^2 does not fall in the rejection region, we cannot conclude that there is a difference in the preference for the five categories.

14.7 One thousand cars were each classified according to the lane which they occupied (one through four). If no lane is preferred over another, the probability that a car will be driven in lane i, $i = 1, 2, 3, 4$ is ¼. The null hypothesis is then

$$H_0 : p_1 = p_2 = p_3 = p_4 = \frac{1}{4}$$

and the test statistic is

$$X^2 = \sum \frac{(O_i - E_i)^2}{E_i}$$

with $E_i = np_i = 1000(1/4) = 250$ for $i = 1, 2, 3, 4$. A table of observed and expected cell counts follows:

Lane	1	2	3	4
O_i	294	276	238	192
E_i	250	250	250	250

Then

$$X^2 = \frac{(294-250)^2}{250} + \frac{(276-250)^2}{250} + \frac{(238-250)^2}{250} + \frac{(192-250)^2}{250}$$

$$= \frac{6120}{250} = 24.48$$

The rejection region with $k - 1 = 3$ df is $X^2 > \chi^2_{.05} = 7.81$. Since the observed value of X^2 falls in the rejection region, we reject H_0. There is a difference in preference for the four lanes (they seem to prefer lane #1).

14.9   If the frequency of occurrence of a heart attack is the same for each day of the week, then when a heart attack occurs, the probability that it falls in one cell (day) is the same as for any other cell (day). Hence,

$$H_0 : p_1 = p_2 = \cdots = p_7 = \frac{1}{7}$$

vs. H_a : at least one p_i is different from the others , or equivalently,

$$H_a : p_i \neq p_j \text{ for some pair } i \neq j$$

Since $n = 200$, $E_i = np_i = 200(1/7) = 28.571429$ and the test statistic is

$$X^2 = \frac{(24 - 28.571429)^2}{28.571429} + \cdots + \frac{(29 - 28.571429)^2}{28.571429} = \frac{103.71429}{28.571429} = 3.63$$

The degrees of freedom for this test of specified cell probabilities is $k - 1 = 7 - 1 = 6$ and the upper tailed rejection region is

$$X^2 > \chi^2_{.05} = 12.59$$

H_0 is not rejected. There is insufficient evidence to indicate a difference in frequency of occurrence from day to day.

14.11   Similar to previous exercises. The hypothesis to be tested is

$$H_0 : p_1 = p_2 = \cdots = p_{12} = \frac{1}{12}$$

versus H_a : at least one p_i is different from the others
with

$$E_i = np_i = 400(1/12) = 33.333 .$$

The test statistic is

$$X^2 = \frac{(38 - 33.33)^2}{33.33} + \cdots + \frac{(35 - 33.33)^2}{33.33} = 13.58$$

The upper tailed rejection region is with $\alpha = .05$ and $k - 1 = 11$ df is $X^2 > \chi^2_{.05} = 19.675$. The null hypothesis is not rejected and we cannot conclude that the proportion of cases varies from month to month.

14.13   It is necessary to determine whether proportions given by the Mars Company are correct. The null hypothesis to be tested is

$$H_0 : p_1 = .13; \ p_2 = .14; \ p_3 = .13; \ p_4 = .24; \ p_5 = .20; \ p_6 = .16$$

184

versus the alternative that at least one of these probabilities is incorrect. A table of observed and expected cell counts follows:

Color	Brown	Yellow	Red	Blue	Orange	Green
O_i	70	72	61	118	108	85
E_i	66.82	71.96	66.82	123.36	102.80	82.24

The test statistic is

$$X^2 = \frac{(70-66.82)^2}{66.82} + \frac{(72-71.96)^2}{71.96} + \cdots + \frac{(85-82.24)^2}{82.24} = 1.247$$

The number of degrees of freedom is $k-1 = 5$ and, since the observed value of $X^2 = 1.247$ is less than $\chi^2_{.10} = 9.24$, the p-value is greater than .10 and the results are not significant. We conclude that the proportions reported by the Mars Company are substantiated by our sample.

14.15 It is necessary to determine whether admission rates differ from the previously reported rates. A table of observed and expected cell counts follows:

	Unconditional	Trial	Refused	Totals
O_i	329	43	128	500
E_i	300	25	175	500

The null hypothesis to be tested is

$$H_0 : p_1 = .60; \ p_2 = .05; \ p_3 = .35$$

against the alternative that at least one of these probabilities is incorrect. The test statistic is

$$X^2 = \frac{(329-300)^2}{300} + \frac{(43-25)^2}{25} + \frac{(128-175)^2}{175} = 28.386$$

The number of degrees of freedom is $k-1 = 2$ and the rejection region $X^2 > \chi^2_{.05} = 5.99$. The null hypothesis is rejected, and we conclude that there has been a departure from previous admission rates. Notice that the percentage of unconditional admissions has risen slightly, the number of conditional admissions has increased, and the percentage refused admission has decreased at the expense of the fist two categories.

14.17 Refer to Section 14.4 of the text. For a 3×5 contingency table with $r = 3$ and $c = 5$, there are $(r-1)(c-1) = (2)(4) = 8$ degrees of freedom.

14.19 The hypothesis to be tested is

H_0 : opinion groups are independent of gender
H_a : opinion groups are dependent on gender

and the test statistic is the chi-square statistic given in the printout as $X^2 = 18.352$ with p-value $= .000$. Because of the small p-value, the results are highly significant, and H_0 is rejected. There is evidence of a dependence between opinion group and gender. The conditional distributions of the three groups for males and females are shown in the following table.

	Group 1	Group 2	Group 3	Total
Males	$\frac{37}{158} = .23$	$\frac{49}{158} = .31$	$\frac{72}{158} = .46$	1.00
Females	$\frac{7}{88} = .08$	$\frac{50}{88} = .57$	$\frac{31}{88} = .35$	1.00

You can see that men tend to favor the group 3 opinion, while almost 60% of the women favor the group 2 opinion. The group 1 opinion contains a very small proportion of the women, but almost 25% of the men.

14.21 **a** The hypothesis of independence between attachment pattern and child care time is tested using the chi-square statistic. The contingency table, including column and row totals and the estimated expected cell counts, follows.

Attachment	Child Care Low	Moderate	High	Total
Secure	24 (24.09)	35 (30.97)	5 (8.95)	64
Anxious	11 (10.91)	10 (14.03)	8 (4.05)	29
Total	111	51	297	459

The test statistic is

$$X^2 = \frac{(24-24.09)^2}{24.09} + \frac{(35-30.97)^2}{30.97} + \cdots + \frac{(8-4.05)^2}{4.05} = 7.267$$

and the rejection region is $X^2 > \chi^2_{.05} = 5.99$ with 2 df. H_0 is rejected. There is evidence of a dependence between attachment pattern and child care time.

b The value $X^2 = 7.267$ is between $\chi^2_{.05}$ and $\chi^2_{.025}$ so that $.025 < p\text{-value} < .05$. The results are significant.

14.23 **a** The hypothesis of no difference in the proportions of men and women in each hair color category is equivalent to testing the hypothesis of independence between hair color and gender, and is tested using the chi-square statistic. The contingency table, including column and row totals and the estimated expected cell counts, follows.

Gender	Hair Color Light Blonde	Blonde	Light Brown	Brown	Black	Red	Total
Male	4 (4.385)	46 (50.667)	45 (55.538)	176 (165.641)	23 (17.051)	10 (10.718)	304
Female	5 (4.615)	58 (53.333)	69 (58.462)	164 (174.359)	12 (17.949)	12 (11.282)	320
Total	9	104	114	340	35	22	624

The test statistic is

$$X^2 = \frac{(4-4.385)^2}{4.385} + \frac{(46-50.667)^2}{50.667} + \cdots + \frac{(12-11.282)^2}{11.282} = 10.208 \ (10.207 \text{ using full accuracy})$$

and the rejection region is $X^2 > \chi^2_{.05} = 11.0705$ with 5 df. Alternatively, we can bound the p-value as

$$.05 < p\text{-value} = P(X^2 > 10.208) < .10$$

and H_0 is not rejected at the 5% level. There is insufficient evidence to indicate a difference in the proportion of individuals with these hair colors between men and women.

b The contingency table is collapsed to combine "light blonde" and "blonde" into one category. The table is shown below, along with the estimated expected cell counts.

Gender	Hair Color Blonde	Light Brown	Brown	Black	Red	Total
Male	50 (55.051)	45 (55.538)	176 (14.14)	23 (17.051)	10 (10.718)	304
Female	63 (57.949)	69 (58.462)	164 (174.359)	12 (17.949)	12 (11.282)	320
Total	113	114	340	35	22	624

The test statistic is

$$X^2 = \frac{(50-55.051)^2}{55.051} + \cdots + \frac{(12-11.282)^2}{11.282} = 10.207$$

and the rejection region is $X^2 > \chi^2_{.05} = 9.48773$ with 4 df. Alternatively, we can bound the p-value as

186

$$.025 < p\text{-value} = P(X^2 > 10.207) < .05$$

and H_0 *is* rejected at the 5% level. There *is* sufficient evidence to indicate a difference in the proportion of individuals with these hair colors between men and women.

14.25 **a** The hypothesis of independence between salary and number of workdays at home is tested using the chi-square statistic. The contingency table, including column and row totals and the estimated expected cell counts, generated by *Minitab* follows.

Chi-Square Test: Less than one, At least one, not all, All at home

```
Expected counts are printed below observed counts
Chi-Square contributions are printed below expected counts

                At
             least
       Less   one,    All
       than   not     at
       one    all     home   Total
   1     38     16     14      68
       36.27  21.08  10.65
       0.083  1.224  1.051

   2     54     26     12      92
       49.07  28.52  14.41
       0.496  0.223  0.404

   3     35     22      9      66
       35.20  20.46  10.34
       0.001  0.116  0.174

   4     33     29     12      74
       39.47  22.94  11.59
       1.060  1.601  0.014

Total   160     93     47     300

Chi-Sq = 6.447, DF = 6, P-Value = 0.375
```

The test statistic is

$$X^2 = \frac{(38-36.27)^2}{36.27} + \frac{(16-21.08)^2}{21.08} + \cdots + \frac{(12-11.59)^2}{11.59} = 6.447$$

and the rejection region with $\alpha = .05$ and $df = 3(2) = 6$ is $X^2 > \chi^2_{.05} = 12.59$ and the null hypothesis is not rejected. There is insufficient evidence to indicate that salary is dependent on the number of workdays spent at home.

b The observed value of the test statistic, $X^2 = 6.447$, is less than $\chi^2_{.10} = 10.6446$ so that the *p*-value is more than .10. This would confirm the non-rejection of the null hypothesis from part **a**.

14.27 Similar to previous exercises, except that the number of observations per row were selected prior to the experiment. The test procedure is identical to that used for an $r \times c$ contingency table. The contingency table, including column and row totals and the estimated expected cell counts, follows.

Population	Category			Total
	1	2	3	
1	108	52	40	200
	(102.33)	(47.33)	(50.33)	
2	87	51	62	200
	(102.33)	(47.33)	(50.33)	
3	112	39	49	200
	(102.33)	(47.33)	(50.33)	
Total	307	142	151	600

a The test statistic is

$$X^2 = \frac{(108-102.33)^2}{102.33} + \frac{(52-47.33)^2}{47.33} + \cdots + \frac{(49-50.33)^2}{50.33} = 10.597$$

using calculator accuracy.

b With $(r-1)(c-1) = 4$ df and $\alpha = .01$, the rejection region is $X^2 > 13.2767$.

c The null hypothesis is not rejected. There is insufficient evidence to indicate that the proportions in each of the three categories depend upon the population from which they are drawn.

d Since the observed value, $X^2 = 10.597$, falls between $\chi^2_{.05}$ and $\chi^2_{.025}$,

$$.025 < p\text{-value} < .05 .$$

14.29 Because a set number of Americans in each sub-population were each fixed at 200, we have a contingency table with fixed rows. The table, with estimated expected cell counts appearing in parentheses, follows.

	Yes	No	Total
White-American	40 (62)	160 (138)	200
African-American	56 (62)	144 (138)	200
Hispanic-American	68 (62)	132 (138)	200
Asian-American	84 (62)	116 (138)	200
Total	248	552	800

The test statistic is

$$X^2 = \frac{(40-62)^2}{62} + \frac{(56-62)^2}{62} + \cdots + \frac{(116-138)^2}{138} = 24.31$$

and the rejection region with 3 df is $X^2 > 11.3449$. H_0 is rejected and we conclude that the incidence of parental support is dependent on the sub-population of Americans.

14.31 **a** Each of the three care type facilities for long-term care represents a binomial population in which we measure the presence or absence of EMI services.

b The 3×2 contingency table is analyzed as in previous exercises. The *Minitab* printout below shows the observed and estimated expected cell counts, the test statistic and its associated *p*-value.

Chi-Square Test: EMI, Non-EMI
```
Expected counts are printed below observed counts
Chi-Square contributions are printed below expected counts
        EMI   Non-EMI   Total
  1      54        22      76
       42.90     33.10
       2.873     3.723
  2      59        77     136
       76.77     59.23
       4.112     5.329
  3      49        26      75
       42.33     32.67
       1.049     1.360
Total   162       125     287
Chi-Sq = 18.446, DF = 2, P-Value = 0.000
```

The results are significant at the 1% level (p-value $= .000$) and we conclude that the type of care provided varies by the three types of long-term care facility.

c Calculate the proportion of homes of the three care types which have EMI services:

$$\hat{p}_N = \frac{54}{76} = .71, \quad \hat{p}_R = \frac{59}{136} = .43, \quad \text{and } \hat{p}_D = \frac{49}{75} = .65.$$

It appears that nursing care and dual-registered facilities are more likely to have EMI services than is residential care.

14.33 The number of observations per column was selected prior to the experiment. The test procedure is identical to that used for an $r \times c$ contingency table. The contingency table, including column and row totals and the estimated expected cell counts, follows.

Family Members	Type			Total
	Apartment	Duplex	Single Residence	
1	8 (9.67)	20 (9.67)	1 (9.67)	29
2	16 (11)	8 (11)	9 (11)	33
3	10 (11.33)	10 (11.33)	14 (11.33)	34
4 or more	6 (8)	2 (8)	16 (8)	24
Total	40	40	40	120

The test statistic is

$$X^2 = \frac{(8-9.67)^2}{9.67} + \frac{(20-9.67)^2}{9.67} + \cdots + \frac{(16-8)^2}{8} = 36.499$$

using computer accuracy. With $(r-1)(c-1) = 6 \ df$ and $\alpha = .01$, the rejection region is $X^2 > 16.8119$. The null hypothesis is rejected. There is sufficient evidence to indicate that family size is dependent on type of family residence. It appears that as the family size increases, it is more likely that people will live in single residences.

14.35 If the housekeeper actually has no preference, he or she has an equal chance of picking any of the five floor polishes. Hence, the null hypothesis to be tested is

$$H_0 : p_1 = p_2 = p_3 = p_4 = p_5 = \frac{1}{5}$$

The values of O_i are the actual counts observed in the experiment, and $E_i = np_i = 100(1/5) = 20$.

Polish	A	B	C	D	E
O_i	27	17	15	22	19
E_i	20	20	20	20	20

Then

$$X^2 = \frac{(27-20)^2}{20} + \frac{(17-20)^2}{20} + \cdots + \frac{(19-20)^2}{20} = 4.40$$

The p-value with $df = k - 1 = 4$ is greater than .10 and H_0 is not rejected. We cannot conclude that there is a difference in the preference for the five floor polishes. Even if this hypothesis **had** been rejected, the conclusion would be that at least one of the values of the p_i was significantly different from 1/6. However, this does not imply that p_i is necessarily greater than 1/6. Hence, we could not conclude that polish A is superior.

If the objective of the experiment is to show that polish A is superior, a better procedure would be to test the hypothesis as follows:

$$H_0 : p_1 = 1/6 \qquad H_a : p_1 > 1/6$$

From a sample of $n = 100$ housewives, $x = 27$ are found to prefer polish A. A z-test can be performed on the single binomial parameter p_1.

14.37 The data is analyzed as a 2×3 contingency table with estimated expected cell counts shown in parentheses.

189

	Small	Medium	Large	Total
Fatal	67 (61.43)	26 (28.04)	16 (19.53)	109
Not Fatal	128 (133.57)	63 (60.96)	46 (42.47)	237
Total	195	89	62	346

The test statistic is

$$X^2 = \frac{(67-61.43)^2}{61.43} + \frac{(26-28.04)^2}{28.04} + \cdots + \frac{(46-42.47)^2}{42.47} = 1.885$$

The p-value with 2 df is greater than .10 and H_0 is not rejected. There is insufficient evidence to indicate that the frequency of fatal accidents is depending on the size of automobiles.

14.39 **a** To test for equality of the two binomial proportions, we use chi-square statistic and the 2×2 contingency table shown below The null hypothesis is evaluation (positive or negative) is independent of teaching approach; that is, there is no difference in p the proportion of positive evaluations for the discovery based versus the standard teaching approach. The contingency table generated by *Minitab* is shown below.

Chi-Square Test: Positive, Negative
```
Expected counts are printed below observed counts
Chi-Square contributions are printed below expected counts

       Positive  Negative  Total
   1         37        11     48
         34.00     14.00
         0.265     0.643

   2         31        17     48
         34.00     14.00
         0.265     0.643

Total        68        28     96
Chi-Sq = 1.815, DF = 1, P-Value = 0.178
```

The observed value of the test statistic is $X^2 = 1.815$ with p-value $= .178$ and the null hypothesis is not rejected at the 5% level of significance. There is insufficient evidence to indicate that there is a difference in the proportion of positive evaluations for the discovery based versus the standard teaching approach.
b Since the observed value of the test statistic is less than $\chi^2_{.10} = 2.71$ with $df = 1$, the approximate p-value is greater than .10. This agrees with the actual p-value $= .178$ given in the printout.

14.41 In Exercise 14.40, we found that the observed value of the test statistic is $z == 3.1297$ so that the one-tailed p-value is

$$p\text{-value} = P(z > 3.13) < (.5 - .4990) = .001$$

The *Minitab* printout shows the chi-square test ($X^2 = 9.795$) with p-value $= .002$. This is not an inconsistent result, if you remember that the chi-square test rejects H_0 in favor of a non-directional alternative using only large values of X^2. To obtain the directional test in Exercise 14.40**a**, you must first check to see that the direction of the difference is as indicated by $H_a(\hat{p}_1 > \hat{p}_2)$. The p-value for the test is then half the two-tailed p-value given by *Minitab* or

$$p\text{-value} = \frac{1}{2}(.002) = .001$$

Also, the two test statistics are related as:

$$z^2 = (3.1297)^2 = 9.795 = X^2$$

14.43 The flower fall into one of four classifications, with theoretical ratio 9:3:3:1. Converting these ratios to probabilities,

$$p_1 = 9/16 = .5625 \qquad p_2 = 3/16 = .1875$$
$$p_3 = 3/16 = .1875 \qquad p_4 = 1/16 = .0625$$

We will test the null hypothesis that the probabilities are as above against the alternative that they differ. The table of observed and expected cell counts follows:

	AB	Ab	aB	aa
O_i	95	30	28	7
E_i	90	30	30	10

The test statistic is

$$X^2 = \frac{(95-90)^2}{90} + \frac{(30-30)^2}{30} + \frac{(28-30)^2}{30} + \frac{(7-10)^2}{10} = 1.311$$

The number of degrees of freedom is $k-1 = 3$ and the rejection region with $\alpha = .01$ is $X^2 > \chi^2_{.01} = 11.3449$. Since the observed value of X^2 does not fall in the rejection region, we do not reject H_0. We do not have enough information to contradict the theoretical model for the classification of flower color and shape.

14.45 **a** Similar to previous exercises. The contingency table, including column and row totals and the estimated expected cell counts, follows.

Condition	Treated	Untreated	Total
Improved	117 (95.5)	74 (95.5)	191
Not improved	83 (104.5)	126 (104.5)	209
Total	200	200	400

The test statistic is

$$X^2 = \frac{(117-95.5)^2}{95.5} + \frac{(74-95.5)^2}{95.5} + \cdots + \frac{(126-104.5)^2}{104.5} = 18.527$$

To test a one-tailed alternative of "effectiveness", first check to see that $\hat{p}_1 > \hat{p}_2$. Then the rejection region with 1 df has a right-tail area of $2(.05) = .10$ or $X^2 > \chi^2_{2(.05)} = 2.706$. H_0 is rejected and we conclude that the serum is effective.

b Consider the treated and untreated patients as comprising random samples of two hundred each, drawn from two populations (i.e., a sample of 200 treated patients and a sample of 200 untreated patients). Let p_1 be the probability that a treated patient improves and let p_2 be the probability that an untreated patient improves. Then the hypothesis to be tested is

$$H_0 : p_1 - p_2 = 0 \qquad H_a : p_1 - p_2 > 0$$

Using the procedure described in Chapter 9 for testing the hypothesis about the difference between two binomial parameters, the following estimators are calculated:

$$\hat{p}_1 = \frac{x_1}{n_1} = \frac{117}{200} \qquad \hat{p}_2 = \frac{x_2}{n_2} = \frac{74}{200} \qquad \hat{p} = \frac{x_1 + x_2}{n_1 + n_2} = \frac{117+74}{400} = .4775$$

The test statistic is

$$z = \frac{\hat{p}_1 - \hat{p}_2 - 0}{\sqrt{\hat{p}\hat{q}(1/n_1 + 1/n_2)}} = \frac{.215}{\sqrt{.4775(.5225)(.01)}} = 4.304$$

And the rejection region for $\alpha = .05$ is $z > 1.645$. Again, the test statistic falls in the rejection region. We reject the null hypothesis of no difference and conclude that the serum is effective. Notice that

$$z^2 = (4.304)^2 = 18.52 = X^2 \text{ (to within rounding error)}$$

14.47 Refer to Section 9.6. The two-tailed z test was used to test the hypothesis

$$H_0 : p_1 - p_2 = 0 \qquad H_a : p_1 - p_2 \neq 0$$

using the test statistic

$$z = \frac{\hat{p}_1 - \hat{p}_2}{\sqrt{\hat{p}\hat{q}\left(\dfrac{1}{n_1} + \dfrac{1}{n_2}\right)}}$$

$$\Rightarrow \quad z = \frac{\left(\hat{p}_1 - \hat{p}_2\right)^2}{\hat{p}\hat{q}\left(\dfrac{n_1 + n_2}{n_1 n_2}\right)} = \frac{n_1 n_2 \left(\hat{p}_1 - \hat{p}_2\right)^2}{\hat{p}\hat{q}\left(n_1 + n_2\right)}$$

Note that

$$\hat{p} = \frac{x_1 + x_2}{n_1 + n_2} = \frac{n_1 \hat{p}_1 + n_2 \hat{p}_2}{n_1 + n_2}$$

Now consider the chi-square test statistic used in Exercise 14.45. The hypothesis to be tested is

H_0 : independence of classification $\qquad$ H_a : dependence of classification

That is, the null hypothesis asserts that the percentage of patients who show improvement is independent of whether or not they have been treated with the serum. If the null hypothesis is true, then $p_1 = p_2$. Hence, the two tests are designed to test the same hypothesis. In order to show that z^2 is equivalent to X^2, it is necessary to rewrite the chi-square test statistic in terms of the quantities, $\hat{p}_1, \hat{p}_2, \hat{p}, n_1$ and n_2.

1 Consider O_{11}, the observed number of treated patients who have improved. Since $\hat{p}_1 = O_{11}/n_1$, we have

$O_{11} = n_1 \hat{p}_1$. Similarly,

$$O_{21} = n_1 \hat{q}_1 \qquad O_{12} = n_2 \hat{p}_2 \qquad O_{22} = n_2 \hat{q}_2$$

2 The estimated expected cell counts are calculated under the assumption that the null hypothesis is true. Consider

$$E_{11} = \frac{r_1 c_1}{n} = \frac{\left(O_{11} + O_{12}\right)\left(O_{11} + O_{21}\right)}{n_1 + n_2} = \frac{\left(x_1 + x_2\right)\left(O_{11} + O_{21}\right)}{n_1 + n_2} = n_1 \hat{p}$$

Similarly,

$$\hat{E}_{21} = n_1 \hat{q} \qquad \hat{E}_{12} = n_2 \hat{p} \qquad \hat{E}_{22} = n_2 \hat{q}$$

The table of observed and estimated expected cell counts follows.

	Treated	Untreated	Total
Improved	$n_1 \hat{p}_1$ $\left(n_1 \hat{p}\right)$	$n_2 \hat{p}_2$ $\left(n_2 \hat{p}\right)$	$x_1 + x_2$
Not improved	$n_1 \hat{q}_1$ $\left(n_1 \hat{q}\right)$	$n_2 \hat{q}_1$ $\left(n_2 \hat{q}\right)$	$n - \left(x_1 + x_2\right)$
Total	n_1	n_2	n

Then

$$X^2 = \sum \frac{\left(O_{ij} - E_{ij}\right)^2}{E_{ij}}$$

$$= \frac{n_1^2 \left(\hat{p}_1 - \hat{p}\right)^2}{n_1 \hat{p}} + \frac{n_1^2 \left(\hat{q}_1 - \hat{q}\right)^2}{n_1 \hat{q}} + \frac{n_2^2 \left(\hat{p}_2 - \hat{p}\right)^2}{n_2 \hat{p}} + \frac{n_2^2 \left(\hat{q}_2 - \hat{q}\right)^2}{n_2 \hat{q}}$$

$$= \frac{n_1 \left(\hat{p}_1 - \hat{p}\right)^2}{\hat{p}} + \frac{n_1 \left[\left(1 - \hat{p}_1\right) - \left(1 - \hat{p}\right)\right]^2}{\hat{q}} + \frac{n_2 \left(\hat{p}_2 - \hat{p}\right)^2}{\hat{p}} + \frac{n_2 \left[\left(1 - \hat{p}_2\right) - \left(1 - \hat{p}\right)\right]^2}{\hat{q}}$$

$$= \frac{\left(1 - \hat{p}_1\right)n_1 \left(\hat{p}_1 - \hat{p}\right)^2 + n_1 \hat{p}\left(\hat{p}_1 - \hat{p}\right)^2}{\hat{p}\hat{q}} + \frac{\left(1 - \hat{p}_2\right)n_2 \left(\hat{p}_2 - \hat{p}\right)^2 + n_2 \hat{p}\left(\hat{p}_2 - \hat{p}\right)^2}{\hat{p}\hat{q}}$$

$$= \frac{n_1 \left(\hat{p}_1 - \hat{p}\right)^2}{\hat{p}\hat{q}} + \frac{n_2 \left(\hat{p}_2 - \hat{p}\right)^2}{\hat{p}\hat{q}}$$

Substituting for $\hat{p}$, we obtain

$$X^2 = \frac{n_1}{\hat{p}\hat{q}}\left[\frac{n_1\hat{p}_1 + n_2\hat{p}_1 - n_1\hat{p}_1 - n_2\hat{p}_2}{n_1 + n_2}\right]^2 + \frac{n_2}{\hat{p}\hat{q}}\left[\frac{n_1\hat{p}_2 + n_2\hat{p}_2 - n_1\hat{p}_1 - n_2\hat{p}_2}{n_1 + n_2}\right]^2$$

$$= \frac{n_1 n_2^2 (\hat{p}_1 - \hat{p}_2)^2 + n_1^2 n_2 (\hat{p}_1 - \hat{p}_2)^2}{\hat{p}\hat{q}(n_1 + n_2)^2} = \frac{n_1 n_2 (\hat{p}_1 - \hat{p}_2)^2}{\hat{p}\hat{q}(n_1 + n_2)}$$

Note that X^2 is identical to z^2, as defined at the beginning of the exercise.

14.49 The null hypothesis is that the two methods of classification are independent. The 2×2 contingency table with estimated expected cell counts in parentheses follows.

	Infection	No Infection	Total
Antibody	4 (8.913)	78 (73.087)	82
No antibody	11 (6.087)	45 (49.913)	56
Total	15	123	138

The test statistic is

$$X^2 = \frac{(4 - 8.913)^2}{8.913} + \frac{(78 - 73.087)^2}{73.087} + \cdots + \frac{(45 - 49.913)^2}{49.913} = 7.487$$

($X^2 = 7.488$ using computer accuracy). The rejection region with $\alpha = .05$ and 1 df is $X^2 > 3.84$ (alternatively, the p-value is bounded as $.005 < p\text{-value} < .01$) and H_0 is rejected. There is evidence that the injection of antibodies affects the likelihood of infections.

14.51 The null hypothesis to be tested is

$$H_0 : p_1 = p_2 = p_3 = \frac{1}{3}$$

and the test statistic is

$$X^2 = \sum \frac{(O_i - E_i)^2}{E_i}$$

with $E_i = np_i = 200(1/3) = 66.67$ for $i = 1, 2, 3$. A table of observed and expected cell counts follows:

Entrance	1	2	3
O_i	83	61	56
E_i	66.67	66.67	66.67

Then

$$X^2 = \frac{(84 - 66.67)^2}{66.67} + \frac{(61 - 66.67)^2}{66.67} + \frac{(56 - 66.67)^2}{66.67} = 6.190$$

With $df = k - 1 = 2$, the p-value is between .025 and .05 and we can reject H_0 at the 5% level of significance. There is a difference in preference for the three doors. A 95% confidence interval for p_1 is given as

$$\frac{x_1}{n} \pm z_{.025}\sqrt{\frac{\hat{p}_1\hat{q}_1}{n}} \Rightarrow \frac{83}{200} \pm 1.96\sqrt{\frac{.415(.585)}{200}} \Rightarrow .415 \pm .068$$

or $.347 < p_1 < .483$.

14.53 a The 2×4 contingency table with observed and estimated expected cell counts is shown below. Remember that $E_{ij} = \frac{(r_i)(c_j)}{n}$.

	Very Confident	Somewhat Confident	Not too Confident	Not at all Confident	Total
Men	210 (164.5)	241 (273.5)	68 (70.5)	5 (10.5)	524
Women	129 (164.5)	306 (273.5)	73 (70.5)	16 (10.5)	524
Total	329	547	141	21	1048

Then the test statistic is

$$X^2 = \frac{(210-164.5)^2}{164.5} + \frac{(241-273.5)^2}{273.5} + \cdots + \frac{(16-10.5)^2}{10.5} = 33.017$$

with $df = (r-1)(c-1) = 3$. With $\alpha = .05$, the rejection region is $X^2 \geq \chi^2_{3,.05} = 7.81$ and H_0 is rejected.

b Since the observed value of $X^2 = 33.017$ is larger than the largest value in Table 5 ($\chi^2_{.005} = 12.8381$), the p-value is less than .005.

14.55 The null hypothesis to be tested is

H_0 : color distribution is independent of vehicle type

H_a : color distribution is dependent on vehicle type

A table of observed and expected cell counts follows, with $E_{ij} = \frac{(r_i)(c_j)}{n}$.

Color	Silver	Black	Gray	Blue	Red	White	Other	Total
Compact/Sports	52 (51)	43 (38)	48 (42.5)	41 (36.5)	32 (29.5)	19 (28.5)	15 (24)	250
Full Intermediate	50 (51)	33 (38)	37 (42.5)	32 (36.5)	27 (29.5)	38 (28.5)	33 (24)	250
Total	102	76	85	73	59	57	48	500

Then the test statistic is

$$X^2 = \frac{(52-51)^2}{51} + \frac{(43-38)^2}{38} + \cdots + \frac{(33-24)^2}{24} = 17.395$$

Since the observed value of X^2 with $df = k - 1 = 6$ is between $\chi^2_{.01}$ and $\chi^2_{.005}$, the p-value is between .005 and .01 (a computer printout will show p-value = .008). We can reject H_0. There is sufficient evidence to suggest a difference in color distributions for the two types of vehicles.

14.57 **a** The *Minitab* printout gives $X^2 = 3.660$ with p-value = .454. The results are not significant; H_0 is not rejected, and there is insufficient evidence to indicate a difference in the distribution of injury types for rugby forwards and backs. However, *Minitab* warns you that the assumption that E_i is greater than or equal to five for each cell has been violated. Since the effect of a small expected value is to inflate the value of X^2, you need not be too concerned (X^2 was not big enough to reject H_0 anyway).
b A difference in the proportion of MCL sprains for the two positions involves a 2×2 contingency table or a z test for the difference in two binomial proportions. The *Minitab* chi-square printout is shown on the next page.

Chi-Square Test: Forwards, Backs

```
Expected counts are printed below observed counts
Chi-Square contributions are printed below expected counts
        Forwards  Backs  Total
    1       14       9     23
            11.50   11.50
            0.543   0.543

    2       24      29     53
            26.50   26.50
            0.236   0.236

Total       38      38     76
Chi-Sq = 1.559, DF = 1, P-Value = 0.212
```

The *p*-value is greater than the .05 claimed by the researcher, and does not suggest a significant difference. The test for the difference in the proportion of ACL tears is shown in the *Minitab* printout with *p*-value = .073 . Again, these results do not agree with the author's calculations.

Chi-Square Test: Forwards, Backs

```
Expected counts are printed below observed counts
Chi-Square contributions are printed below expected counts
        Forwards  Backs  Total
    1        7      14     21
            10.50   10.50
            1.167   1.167

    2       31      24     55
            27.50   27.50
            0.445   0.445

Total       38      38     76
Chi-Sq = 3.224, DF = 1, P-Value = 0.073
```

14.59   **a**   The 2×3 contingency table is analyzed as in previous exercises. The *Minitab* printout below shows the observed and estimated expected cell counts, the test statistic and its associated *p*-value.

Chi-Square Test: Three or fewer, four or Five, Six or More

```
Expected counts are printed below observed counts
Chi-Square contributions are printed below expected counts

        Three or  four or  Six or
         fewer     Five     More   Total
    1       49       43       34     126
            36.52    45.65    43.83
            4.263    0.154    2.203

    2       31       57       62     150
            43.48    54.35    52.17
            3.581    0.129    1.851

Total       80      100       96     276
Chi-Sq = 12.182, DF = 2, P-Value = 0.002
```

The results are highly significant (*p*-value = .002) and we conclude that there is a difference in the susceptibility to colds depending on the number of relationships you have.

b   The proportion of people with colds is calculated conditionally for each of the three groups, and is shown in the table below.

	Three or fewer	**Four or five**	**Six or more**
Cold	$\dfrac{49}{80} = .61$	$\dfrac{43}{100} = .43$	$\dfrac{34}{96} = .35$
No cold	$\dfrac{31}{80} = .39$	$\dfrac{57}{100} = .57$	$\dfrac{62}{96} = .65$
Total	1.00	1.00	1.00

As the researcher suspects, the susceptibility to a cold seems to decrease as the number of relationships increases!

14.61 The null hypothesis to be tested is

$$H_0 : p_1 = \frac{1}{8}; \ p_2 = \frac{1}{8}; \ p_3 = \frac{1}{8}; \ p_4 = \frac{1}{8}; \ p_5 = \frac{2}{8}; \ p_6 = \frac{2}{8}$$

against the alternative that at least one of these probabilities is incorrect. A table of observed and expected cell counts follows:

Day	Monday	Tuesday	Wednesday	Thursday	Friday	Saturday
O_i	95	110	125	75	181	214
E_i	100	100	100	100	200	200

The test statistic is

$$X^2 = \frac{(95-100)^2}{100} + \frac{(110-100)^2}{100} + \cdots + \frac{(214-200)^2}{200} = 16.535$$

The number of degrees of freedom is $k-1 = 5$ and the rejection region with $\alpha = .05$ is $X^2 > \chi^2_{.05} = 11.07$ and H_0 is rejected. The manager's claim is refuted.

196

15: Nonparametric Statistics

15.1 **a** If distribution 1 is shifted to the right of distribution 2, the rank sum for sample 1 (T_1) will tend to be large. The test statistic will be T_1^*, the rank sum for sample 1 if the observations had been ranked from large to small. The null hypothesis will be rejected if T_1^* is unusually small.

b From Table 7a with $n_1 = 6$, $n_2 = 8$ and $\alpha = .05$, H_0 will be rejected if $T_1^* \le 31$.

c From Table 7c with $n_1 = 6$, $n_2 = 8$ and $\alpha = .01$, H_0 will be rejected if $T_1^* \le 27$.

15.3 **a** H_0 : populations 1 and 2 are identical
H_a : population 1 is shifted to the left of population 2

b The data, with ranks in parentheses, are given below.

Sample 1	Sample 2
1(1)	4(5)
3(3.5)	7(9)
2(2)	6(7.5)
3(3.5)	8(10)
5(6)	6(7.5)

Note that tied observations are given an average rank, the average of the ranks they would have received if they had not been tied. Then
$$T_1 = 1 + 3.5 + 2 + 3.5 + 6 = 16$$
$$T_1^* = n_1(n_1 + n_2 + 1) - T_1 = 5(10 + 1) - 16 = 39$$

c With $n_1 = n_2 = 5$, the one-tailed rejection region with $\alpha = .05$ is found in Table 7a to be $T_1 \le 19$.

d The observed value, $T_1 = 16$, falls in the rejection region and H_0 is rejected. We conclude that population 1 is shifted to the right of population 2.

15.5 If H_a is true and population 1 lies to the right of population 2, then T_1 will be large and T_1^* will be small. Hence, the test statistic will be T_1^* and the large sample approximation can be used. Calculate
$$T_1^* = n_1(n_1 + n_2 + 1) - T_1 = 12(27) - 193 = 131$$
$$\mu_T = \frac{n_1(n_1 + n_2 + 1)}{2} = \frac{12(26 + 1)}{2} = 162$$
$$\sigma_T^2 = \frac{n_1 n_2(n_1 + n_2 + 1)}{12} = \frac{12(14)(27)}{12} = 378$$

The test statistic is
$$z = \frac{T_1 - \mu_T}{\sigma_T} = \frac{131 - 162}{\sqrt{378}} = -1.59$$

The rejection region with $\alpha = .05$ is $z < -1.645$ and H_0 is not rejected. There is insufficient evidence to indicate a difference in the two population distributions.

15.7 The hypothesis of interest is

H_0: populations 1 and 2 are identical   versus H_a: population 2 is shifted to the right of population 1

The data, with ranks in parentheses, are given below.

20s	11(20)	7(11)	6(7.5)	8(14)	6(7.5)	9(16.5)	2(2)	10(18.5)	3(3.5)	6(7.5)
65-70s	1(1)	9(16.5)	6(7.5)	8(14)	7(11)	8(14)	5(5)	7(11)	10(18.5)	3(3.5)

Then

$$T_1 = 20 + 11 + \cdots + 7.5 = 108$$
$$T_1^* = n_1(n_1 + n_2 + 1) - T_1 = 10(20 + 1) - 108 = 102$$

The test statistic is

$$T = \min(T_1, T_1^*) = 102$$

With $n_1 = n_2 = 10$, the one-tailed rejection region with $\alpha = .05$ is found in Table 7a to be $T_1^* \leq 82$ and the observed value, $T = 102$, does not fall in the rejection region; H_0 is not rejected. We cannot conclude that this drug improves memory in mean aged 65 to 70 to that of 20 year olds.

15.9 Similar to previous exercises. The data, with corresponding ranks, are shown in the following table.

Deaf (1)	Hearing (2)
2.75 (15)	0.89 (1)
2.14 (11)	1.43 (7)
3.23 (18)	1.06 (4)
2.07 (10)	1.01 (3)
2.49 (14)	0.94 (2)
2.18 (12)	1.79 (8)
3.16 (17)	1.12 (5.5)
2.93 (16)	2.01 (9)
2.20 (13)	1.12 (5.5)
$T_1 = 126$	

Calculate

$$T_1 = 126$$
$$T_1^* = n_1(n_1 + n_2 + 1) - T_1 = 9(19) - 126 = 45$$

The test statistic is

$$T = \min(T_1, T_1^*) = 45$$

With $n_1 = n_2 = 9$, the two-tailed rejection region with $\alpha = .05$ is found in Table 7b to be $T_1^* \leq 62$. The observed value, $T = 45$, falls in the rejection region and H_0 is rejected. We conclude that the deaf children do differ from the hearing children in eye-movement rate.

15.11 The data, with corresponding ranks, are shown in the following table.

Lake 2	Lake 1
14.1 (12.5)	12.2 (2)
15.2 (16)	13.0 (6)
13.9 (10)	14.1 (12.5)
14.5 (14)	13.6 (7)
14.7 (15)	12.4 (3)
13.8 (8.5)	11.9 (1)
14.0 (11)	12.5 (4)
16.1 (18)	13.8 (8.5)
12.7 (5)	
15.3 (17)	
	$T_1 = 44$

Calculate

$$T_1 = 44$$
$$T_1^* = n_1(n_1 + n_2 + 1) - T_1 = 8(18 + 1) - 44 = 108$$

The test statistic is

$$T = \min(T_1, T_1^*) = 44$$

198

With $n_1 = 8$ and $n_2 = 10$, the two-tailed rejection region with $\alpha = .05$ is found in Table 7b to be $T \leq 53$. The observed value, $T = 44$, falls in the rejection region and H_0 is rejected. We conclude that the distribution of weights for the tagged turtles exposed to the two lake environments were different.

15.13 **a** If a paired difference experiment has been used and the sign test is one-tailed $(H_a : p > .5)$, then the experimenter would like to show that one population of measurements lies above the other population. An exact practical statement of the alternative hypothesis would depend on the experimental situation.
b It is necessary that α (the probability of rejecting the null hypothesis when it is true) take values less than $\alpha = .15$. Assuming the null hypothesis to be true, the two populations are identical and consequently, $p = P(A \text{ exceeds } B \text{ for a given pair of observations})$ is 1/2. The binomial probability was discussed in Chapter 5. In particular, it was noted that the distribution of the random variable x is symmetrical about the mean np when $p = 1/2$. For example, with $n = 25$, $P(x = 0) = P(x = 25)$. Similarly,

$P(x = 1) = P(x = 24)$ and so on. Hence, the lower tailed probabilities tabulated in Table 1, Appendix I will be identical to their upper tailed equivalent probabilities. The values of α available for this upper tailed test and the corresponding rejection regions are shown below.

Rejection Region	α
$x \geq 20$	.002
$x \geq 19$	.007
$x \geq 18$	.022
$x \geq 17$	.054
$x \geq 16$	.115

15.15 Similar to Exercise 15.13. The rejection regions and levels of α are given in the table for the three different values of n, and a one-tailed test.

$n = 10$	$n = 15$	$n = 20$
$x \leq 0$ $\alpha = .001$	$x \leq 2$ $\alpha = .004$	$x \leq 3$ $\alpha = .001$
$x \leq 1$ $\alpha = .011$	$x \leq 3$ $\alpha = .018$	$x \leq 4$ $\alpha = .006$
$x \leq 2$ $\alpha = .055$	$x \leq 4$ $\alpha = .059$	$x \leq 5$ $\alpha = .021$
		$x \leq 6$ $\alpha = .058$
		$x \leq 7$ $\alpha = .132$

For the two-tailed test, the rejection regions with $.01 < \alpha < .15$ are shown below.

$n = 10$	$n = 15$	$n = 20$
$x \leq 0; x \geq 10$ $\alpha = .002$	$x \leq 2; x \geq 13$ $\alpha = .008$	$x \leq 3; x \geq 17$ $\alpha = .002$
$x \leq 1; x \geq 9$ $\alpha = .022$	$x \leq 3; x \geq 12$ $\alpha = .036$	$x \leq 4; x \geq 16$ $\alpha = .012$
$x \leq 2; x \geq 8$ $\alpha = .110$	$x \leq 4; x \geq 11$ $\alpha = .118$	$x \leq 5; x \geq 15$ $\alpha = .042$
		$x \leq 6; x \geq 14$ $\alpha = .116$

15.17 **a** If assessors A and B are equal in their property assessments, then p, the probability that A's assessment exceeds B's assessment for a given property, should equal 1/2. If one of the assessors tends to be more conservative than the other, then either $p > 1/2$ or $p < 1/2$. Hence, we can test the equivalence of the two assessors by testing the hypothesis

$$H_0 : p = 1/2 \quad \text{versus} \quad H_a : p \neq 1/2$$

using the test statistic x, the number of times that assessor A exceeds assessor B for a particular property assessment. To find a two-tailed rejection region with α close to .05, use Table 1 with $n = 8$ and $p = .5$. For the rejection region $\{x = 0, x = 8\}$ the value of α is $.004 + .004 = .008$, while for the rejection region

$\{x = 0,1,7,8\}$ the value of α is $.035 + .035 = .070$ which is closer to $.05$. Hence, using the rejection region $\{x \le 1 \text{ or } x \ge 7\}$, the null hypothesis is not rejected, since $x =$ number of properties for which A exceeds $B = 6$. The p-value for this two-tailed test is

$$p\text{-value} = 2P(x \ge 6) = 2(1 - .855) = .290$$

Since the p-value is greater than $.10$, the results are not significant; H_0 is not rejected (as with the critical value approach).

b The t statistic used in Exercise 10.46 allows the experimenter to reject H_0, while the sign test fails to reject H_0. This is because the sign test used less information and makes fewer assumptions than does the t test. If all normality assumptions are met, the t test is the more powerful test and can reject when the sign test cannot.

15.19   The hypothesis to be tested is

$$H_0 : p = 1/2 \quad \text{versus} \quad H_a : p > 1/2$$

using the sign test with x, the number of "elevated" blood lead levels observed in $n = 17$ people, as the test statistic. Using the large sample approximation, the test statistic is

$$z = \frac{x - .5n}{.5\sqrt{n}} = \frac{15 - .5(17)}{.5\sqrt{17}} = 3.15$$

and the one-tailed rejection region with $\alpha = .05$ is $z > 1.645$. The null hypothesis is rejected and we conclude that the indoor firing range has the effect of increasing a person's blood lead level.

15.21   **a** H_0: population distributions 1 and 2 are identical
H_a: the distributions differ in location

b   Since Table 8, Appendix I gives critical values for rejection in the lower tail of the distribution, we use the smaller of T^+ and T^- as the test statistic.

c   From Table 8 with $n = 30$, $\alpha = .05$ and a two-tailed test, the rejection region is $T \le 137$.

d   Since $T^+ = 249$, we can calculate

$$T^- = \frac{n(n+1)}{2} - T^+ = \frac{30(31)}{2} - 249 = 216.$$

The test statistic is the smaller of T^+ and T^- or $T = 216$ and H_0 is not rejected. There is no evidence of a difference between the two distributions.

15.23   Since $n > 25$, the large sample approximation to the signed rank test can be used to test the hypothesis given in Exercise 15.21a. Calculate

$$E(T) = \frac{n(n+1)}{4} = \frac{30(31)}{4} = 232.5$$

$$\sigma_T^2 = \frac{n(n+1)(2n+1)}{24} = \frac{30(31)(61)}{24} = 2363.75$$

The test statistic is

$$z = \frac{T - E(T)}{\sigma_T} = \frac{216 - 232.5}{\sqrt{2363.75}} = -.34$$

The two-tailed rejection region with $\alpha = .05$ is $|z| > 1.96$ and H_0 is not rejected. The results agree with Exercise 15.21d.

15.25   **a**   The hypothesis to be tested is

H_0: population distributions 1 and 2 are identical
H_a: the distributions differ in location

and the test statistic is T, the rank sum of the positive (or negative) differences. The ranks are obtained by ordering the differences according to their absolute value. Define d_i to be the difference between a pair in populations 1 and 2 (i.e., $x_{1i} - x_{2i}$). The differences, along with their ranks (according to absolute magnitude), are shown in the following table.

d_i	.1	.7	.3	−.1	.5	.2	.5
Rank $| d_i |$	1.5	7	4	1.5	5.5	3	5.5

The rank sum for positive differences is $T^+ = 26.5$ and the rank sum for negative differences is $T^- = 1.5$ with $n = 7$. Consider the smaller rank sum and determine the appropriate lower portion of the two-tailed rejection region. Indexing $n = 7$ and $\alpha = .05$ in Table 8, the rejection region is $T \le 2$ and H_0 is rejected. There is a difference in the two population locations.

b The results do not agree with those obtained in Exercise 15.16. We are able to reject H_0 with the more powerful Wilcoxon test.

15.27 **a** The Wilcoxon signed rank test is used, and the differences, along with their ranks (according to absolute magnitude), are shown in the following table.

d_i	−4	2	−2	−5	−3	0	1	1	−6
Rank $| d_i |$	6	3.5	3.5	7	5	--	1.5	1.5	8

The sixth pair is tied and is hence eliminated from consideration. Pairs 7 and 8, 2 and 3 are tied and receive an average rank. Then $T^+ = 6.5$ and $T^- = 29.5$ with $n = 8$. Indexing $n = 8$ and $\alpha = .05$ in Table 8, the lower portion of the two-tailed rejection region is $T \le 4$ and H_0 is not rejected. There is insufficient evidence to detect a difference in the two machines.

b If a machine continually breaks down, it will eventually be fixed, and the breakdown rate for the following month will decrease.

15.29 **a** The paired data are given in the exercise. The differences, along with their ranks (according to absolute magnitude), are shown in the following table.

d_i	1	2	−1	1	3	1	−1	3	−2	3	1	0
Rank $| d_i |$	3.5	7.5	3.5	3.5	10	3.5	3.5	10	7.5	10	2.5	--

Let $p = P(\text{A exceeds B for a given intersection})$ and $x =$ number of intersections at which A exceeds B. The hypothesis to be tested is

$$H_0 : p = 1/2 \quad \text{versus} \quad H_a : p \ne 1/2$$

using the sign test with x as the test statistic.

Critical value approach: Various two tailed rejection regions are tried in order to find a region with $\alpha \approx .05$. These are shown in the following table.

Rejection Region	α
$x \le 1; x \ge 10$	.012
$x \le 2; x \ge 9$	.066
$x \le 3; x \ge 8$	.226

We choose to reject H_0 if $x \le 2$ or $x \ge 9$ with $\alpha = .066$. Since $x = 8$, H_0 is not rejected. There is insufficient evidence to indicate a difference between the two methods.

p-value approach: For the observed value $x = 8$, calculate the two-tailed p-value:

$$p\text{-value} = 2P(x \ge 8) = 2(1 - .887) = .226$$

Since the p-value is greater than .10, H_0 is not rejected.

b To use the Wilcoxon signed rank test, we use the ranks of the absolute differences shown in the table above. Then $T^+ = 51.5$ and $T^- = 14.5$ with $n = 11$. Indexing $n = 11$ and $\alpha = .05$ in Table 8, the lower portion of the two-tailed rejection region is $T \le 11$ and H_0 is not rejected, as in part **a**.

15.31 **a** Since the experiment has been designed as a paired experiment, there are three tests available for testing the differences in the distributions with and without imagery – (1) the paired difference t test; (2) the sign test or (3) the Wilcoxon signed rank test. In order to use the paired difference t test, the scores must be approximately normal; since the number of words recalled has a binomial distribution with $n = 25$ and unknown recall probability, this distribution may not be approximately normal.

b Using the **sign test**, the hypothesis to be tested is
$$H_0 : p = 1/2 \quad \text{versus} \quad H_a : p > 1/2$$
For the observed value $x = 0$ we calculate the two-tailed p-value:
$$p\text{-value} = 2P(x \le 0) = 2(.000) = .000$$

The results are highly significant; H_0 is rejected and we conclude there is a difference in the recall scores with and without imagery.

Using the **Wilcoxon signed-rank test**, the differences will all be positive ($x = 0$ for the sign test), so that

and $\qquad T^+ = \dfrac{n(n+1)}{2} = \dfrac{20(21)}{2} = 210 \quad$ and $\quad T^- = 210 - 210 = 0$

Indexing $n = 20$ and $\alpha = .01$ in Table 8, the lower portion of the two-tailed rejection region is $T \le 37$ and H_0 is rejected.

15.33 The Kruskal-Wallis H test provides a nonparametric analog to the analysis of variance F test for a completely randomized design presented in Chapter 11. The data are jointly ranked from smallest to largest, with ties treated as in the Wilcoxon rank sum test. The data with corresponding ranks in parentheses are shown below.

	Treatment		
1	2	3	4
124 (9)	147 (20)	141 (17)	117 (4.5)
167 (26)	121 (7)	144 (18.5)	128 (10.5)
135 (14)	136 (15)	139 (16)	102 (1)
160 (24)	114 (3)	162 (25)	119 (6)
159 (23)	129 (12)	155 (22)	128 (10.5)
144 (18.5)	117 (4.5)	150 (21)	123 (8)
133 (13)	109 (2)		
$T_1 = 127.5$	$T_2 = 63.5$	$T_3 = 119.5$	$T_4 = 40.5$
$n_1 = 7$	$n_2 = 7$	$n_3 = 6$	$n_4 = 6$

The test statistic, based on the rank sums, is

$$H = \frac{12}{n(n+1)} \sum \frac{T_i^2}{n_i} - 3(n+1)$$

$$= \frac{12}{26(27)} \left[\frac{(127.5)^2}{7} + \frac{(63.5)^2}{7} + \frac{(119.5)^2}{6} + \frac{(40.5)^2}{6} \right] - 3(27) = 13.90$$

The rejection region with $\alpha = .05$ and $k - 1 = 3$ df is based on the chi-square distribution, or $H > \chi_{.05}^2 = 7.81$. The null hypothesis is rejected and we conclude that there is a difference among the four treatments.

15.35 Similar to Exercise 15.33. The data with corresponding ranks in parentheses are shown below.

	Age		
10 – 19	20 – 39	40 – 59	60 – 69
29 (21)	24 (8)	37 (39)	28 (18)
33 (29.5)	27 (15)	25 (10.5)	29 (21)
26 (12.5)	33 (29.5)	22 (5.5)	34 (34)
27 (15)	31 (24)	33 (29.5)	36 (37.5)
39 (40)	21 (3)	28 (18)	21 (3)
35 (36)	28 (18)	26 (12.5)	20 (1)
33 (29.5)	24 (8)	30 (23)	25 (10.5)
29 (21)	34 (34)	34 (34)	24 (8)
36 (37.5)	21 (3)	27 (15)	33 (29.5)
22 (5.5)	32 (25.5)	33 (29.5)	32 (25.5)
$T_1 = 247.5$	$T_2 = 168$	$T_3 = 216.5$	$T_4 = 188$
$n_1 = 10$	$n_2 = 10$	$n_3 = 10$	$n_4 = 10$

a The test statistic, based on the rank sums, is

$$H = \frac{12}{n(n+1)}\sum\frac{T_i^2}{n_i} - 3(n+1)$$

$$= \frac{12}{40(41)}\left[\frac{(247.5)^2}{10} + \frac{(168)^2}{10} + \frac{(216.5)^2}{10} + \frac{(188)^2}{10}\right] - 3(41) = 2.63$$

The rejection region with $\alpha = .01$ and $k - 1 = 3$ df is based on the chi-square distribution,
or $H > \chi_{.01}^2 = 11.35$. The null hypothesis is not rejected. There is no evidence of a difference in location.

b Since the observed value $H = 2.63$ is less than $\chi_{.10}^2 = 6.25$, the p-value is greater than .10.

c-d From Exercise 11.60, $F = .87$ with 3 and 36 df. Again, the p-value is greater than .10 and the results are the same.

15.37 Similar to previous exercises. The ranks of the data are shown below.

Campaigns		
1	2	3
11.5	6	1.5
7	15	8
1.5	13	4
10	14	11.5
3	5	9
$T_1 = 33$	$T_2 = 53$	$T_3 = 34$
$n_1 = 5$	$n_2 = 5$	$n_3 = 5$

a The test statistic is

$$H = \frac{12}{n(n+1)}\sum\frac{T_i^2}{n_i} - 3(n+1)$$

$$= \frac{12}{15(16)}\left[\frac{(33)^2}{5} + \frac{(53)^2}{5} + \frac{(34)^2}{5}\right] - 3(16) = 2.54$$

With $k - 1 = 2$ df, the observed value $H = 2.54$ is less than $\chi_{.10}^2 = 4.61$, the p-value is greater than .10. The null hypothesis is not rejected and we cannot conclude that there is a difference in the three population distributions.

15.39 In using the Friedman F_r test, data are ranked **within a block** from 1 to k. The treatment rank sums are then calculated as usual. The ranks of the data are shown below.

	Treatment			
Block	1	2	3	4
1	4	1	2	3
2	4	1.5	1.5	3
3	4	1	3	2
4	4	1	2	3
5	4	1	2.5	2.5
6	4	1	2	3
7	4	1	3	2
8	4	1	2	3
	$T_1 = 32$	$T_2 = 8.5$	$T_3 = 18$	$T_4 = 21.5$

a The test statistic is

$$F_r = \frac{12}{bk(k+1)}\sum T_i^2 - 3b(k+1)$$

$$= \frac{12}{8(4)(5)}\left[(32)^2 + (8.5)^2 + 18^2 + (21.5)^2\right] - 3(8)(5) = 21.19$$

and the rejection region is $F_r > \chi^2_{.05} = 7.81$. Hence, H_0 is rejected and we conclude that there is a difference among the four treatments.

b The observed value, $F_r = 21.19$, exceeds $\chi^2_{.005}$, p-value $< .005$.

c-f The analysis of variance is performed as in Chapter 11. The ANOVA table is shown below.

Source	df	SS	MS	F
Treatments	3	198.34375	66.114583	75.43
Blocks	7	220.46875	31.495536	
Error	21	18.40625	0.876488	
Total	31	437.40625		

The analysis of variance F test for treatments is $F = 75.43$ and the approximate p-value with 3 and 21 df is p-value $< .005$. The result is identical to the parametric result.

15.41 Similar to Exercise 15.39, with rats as blocks. The data are shown below along with corresponding ranks within blocks. Note that we have rearranged the data to eliminate the random order of presentation in the display.

		Treatment		
Rat	A	B	C
1	6 (3)	5 (2)	3 (1)
2	9 (2.5)	9 (2.5)	4 (1)
3	6 (2)	9 (3)	3 (1)
4	5 (1)	8 (3)	6 (2)
5	7 (1)	8 (2.5)	8 (2.5)
6	5 (1.5)	7 (3)	5 (1)
7	6 (2)	7 (3)	5 (1)
8	6 (1)	7 (2.5)	7 (2.5)
	$T_1 = 14$	$T_2 = 21.5$	$T_3 = 12.5$

a The test statistic is

$$F_r = \frac{12}{bk(k+1)}\sum T_i^2 - 3b(k+1)$$

$$= \frac{12}{8(3)(4)}\left[(14)^2 + (21.5)^2 + (12.5)^2\right] - 3(8)(4) = 5.81$$

and the rejection region is $F_r > \chi^2_{.05} = 5.99$. Hence, H_0 is not rejected and we cannot conclude that there is a difference among the three treatments.

b The observed value, $F_r = 5.81$, falls between $\chi^2_{.05}$ and $\chi^2_{.10}$. Hence, $.05 < p$-value $< .10$.

15.43 Table 9, Appendix I gives critical values r_0 such that $P(r_s \geq r_0) = \alpha$. Hence, for an upper-tailed test, the critical value for rejection can be read directly from the table.

a $r_s \geq .425$ **b** $r_s \geq .601$

15.45 For a two-tailed test of correlation, the value of α given along the top of the table is doubled to obtain the **actual** value of α for the test.

a To obtain $\alpha = .05$, index .025 and the rejection region is $|r_s| \geq .400$.

b To obtain $\alpha = .01$, index .005 and the rejection region is $|r_s| \geq .526$.

15.47 **a** The two variables (rating and distance) are ranked from low to high, and the results are shown in the following table.

Voter	x	y	Voter	x	y
1	7.5	3	7	6	4
2	4	7	8	11	2
3	3	12	9	1	10
4	12	1	10	5	9
5	10	8	11	9	5.5
6	7.5	11	12	2	5.5

Calculate $\quad \sum x_i y_i = 442.5 \qquad \sum x_i^2 = 649.5 \qquad \sum y_i^2 = 649.5$

$\qquad\qquad n = 12 \qquad\qquad \sum x_i = 78 \qquad\qquad \sum y_i = 78$

Then

$$S_{xy} = 422.5 - \frac{78^2}{12} = -84.5 \qquad S_{xx} = 649.5 - \frac{78^2}{12} = 142.5 \qquad S_{yy} = 649.5 - \frac{78^2}{12} = 142.5$$

and

$$r_s = \frac{S_{xy}}{\sqrt{S_{xx}S_{yy}}} = \frac{-84.5}{142.5} = -.593 \,.$$

b The hypothesis of interest is H_0: no correlation versus H_a: negative correlation. Consulting Table 9 for $\alpha = .05$, the critical value of r_s, denoted by r_0 is $-.497$. Since the value of the test statistic is less than the critical value, the null hypothesis is rejected. There is evidence of a significant negative correlation between rating and distance.

15.49 **a** The data are ranked separately according to the variables x and y.

Rank x	7	6	5	4	1	12	8	3	2	11	10	9
Rank y	7	8	4	5	2	10	12	3	1	6	11	9

Since there were no tied observations, the simpler formula for r_s is used, and

$$r_s = 1 - \frac{6\sum d_i^2}{n(n^2-1)} = 1 - \frac{6\left[(0)^2 + (-2)^2 + \cdots + (0)^2\right]}{12(143)}$$

$$= 1 - \frac{6(54)}{1716} = .811$$

b To test for positive correlation with $\alpha = .05$, index .05 in Table 9 and the rejection region is $r_s \geq .497$. Hence, H_0 is rejected, there is a positive correlation between x and y.

15.51 Refer to Exercise 15.50, with $r_s = .8667$. To test for positive correlation with $\alpha = .05$, index .05 in Table 9 and the rejection region is $r_s \geq .600$. We reject the null hypothesis of no association and conclude that a positive correlation exists between the teacher's ranks and the ranks of the IQs.

15.53 The ranks of the two variables are shown below.

Leaf	1	2	3	4	5	6	7	8	9	10	11	12
Rank x	10.5	5.5	7.5	7.5	4	9	2	5.5	1	12	10.5	3
Rank y	12	7.5	9	6	4.5	10	3	4.5	1	11	7.5	2

Calculate $\quad \sum x_i y_i = 636.25 \qquad \sum x_i^2 = 648.5 \qquad \sum y_i^2 = 649$

$\qquad\qquad n = 12 \qquad\qquad \sum x_i = 78 \qquad\qquad \sum y_i = 78$

Then

$$S_{xy} = 636.25 - \frac{78^2}{12} = 129.25 \qquad S_{xx} = 648.5 - \frac{78^2}{12} = 141.5 \qquad S_{yy} = 649 - \frac{78^2}{12} = 142$$

and

205

$$r_s = \frac{S_{xy}}{\sqrt{S_{xx}S_{yy}}} = \frac{129.25}{\sqrt{141.5(142)}} = .9118 .$$

To test for correlation with $\alpha = .05$, index .025 in Table 9, and the rejection region is $|r_s| \geq .591$. The null hypothesis is rejected and we conclude that there is a correlation between the two variables.

15.55   **a**   Define $p = P(\text{response for stimulus 1 exceeds that for stimulus 2})$ and x = number of times the response for stimulus 1 exceeds that for stimulus 2. The hypothesis to be tested is

$$H_0 : p = 1/2 \quad \text{versus} \quad H_a : p \neq 1/2$$

using the sign test with x as the test statistic. Notice that for this exercise, $n = 9$, and the observed value of the test statistic is $x = 2$. Various two tailed rejection regions are tried in order to find a region with $\alpha \approx .05$. These are shown in the following table.

Rejection Region	α
$x = 0; x = 9$	.004
$x \leq 1; x \geq 8$	.040
$x \leq 2; x \geq 7$	.180

We choose to reject H_0 if $x \leq 1$ or $x \geq 8$ with $\alpha = .040$. Since $x = 2$, H_0 is not rejected. There is insufficient evidence to indicate a difference between the two stimuli.

b   The experiment has been designed in a paired manner, and the paired difference test is used. The differences are shown below.

$$d_i \quad -.9 \quad -1.1 \quad 1.5 \quad -2.6 \quad -1.8 \quad -2.9 \quad -2.5 \quad 2.5 \quad -1.4$$

The hypothesis to be tested is

$$H_0 : \mu_1 - \mu_2 = 0 \qquad H_a : \mu_1 - \mu_2 \neq 0$$

Calculate

$$\bar{d} = \frac{\sum d_i}{n} = \frac{-9.2}{9} = -1.022 \qquad s_d^2 = \frac{\sum d_i^2 - \frac{(\sum d_i)^2}{n}}{n-1} = \frac{37.14 - 9.404}{8} = 3.467$$

and the test statistic is

$$t = \frac{\bar{d}}{\sqrt{\frac{s_d^2}{n}}} = \frac{-1.022}{\sqrt{\frac{3.467}{9}}} = -1.646$$

The rejection region with $\alpha = .05$ and 8 df is $|t| > 2.306$ and H_0 is not rejected.

15.57   **a**   Define $p = P(\text{school A exceeds school B in test score for a pair of twins})$ and x = number of times the score for school A exceeds the score for school B. The hypothesis to be tested is

$$H_0 : p = 1/2 \quad \text{versus} \quad H_a : p \neq 1/2$$

using the sign test with x as the test statistic. Notice that for this exercise, $n = 10$, and the observed value of the test statistic is $x = 7$.

Critical value approach: Various two tailed rejection regions are tried in order to find a region with $\alpha \approx .05$. These are shown in the following table.

Rejection Region	α
$x = 0; x = 10$	.002
$x \leq 1; x \geq 9$	.022
$x \leq 2; x \geq 8$	.110

We choose to reject H_0 if $x \leq 1$ or $x \geq 9$ with $\alpha = .022$. Since $x = 7$, H_0 is not rejected. There is insufficient evidence to indicate a difference between the two schools.

_p_-value approach: For the observed value $x = 7$, calculate the two-tailed p-value:

$$p\text{-value} = 2P(x \geq 7) = 2(1 - .828) = .344$$

and H_0 is not rejected. There is insufficient evidence to indicate a difference between the two schools.

b Consider the one-tailed test of hypothesis as follows:

$$H_0 : p = 1/2 \quad \text{versus} \quad H_a : p > 1/2$$

This alternative will imply that school A is superior to school B. From Table 1, the one-tailed rejection region with $\alpha \approx .05$ is $x \geq 8$ with $\alpha = .055$. The null hypothesis is still not rejected, since $x = 7$. (The one-tailed p-value $= .172$.)

15.59 The data, with corresponding ranks, are shown in the following table.

A (1)	B (2)
6.1 (1)	9.1 (16)
9.2 (17)	8.2 (8)
8.7 (12)	8.6 (11)
8.9 (13.5)	6.9 (2)
7.6 (5)	7.5 (4)
7.1 (3)	7.9 (7)
9.5 (18)	8.3 (9.5)
8.3 (9.5)	7.8 (6)
9.0 (1.5)	8.9(13.5)
$T_1 = 94$	

The difference in the brightness levels using the two processes can be tested using the nonparametric Wilcoxon rank sum test, or the parametric two-sample t test.

1 To test the null hypothesis that the two population distributions are identical, calculate

$$T_1 = 1 + 17 + \cdots + 1.5 = 94$$

$$T_1^* = n_1(n_1 + n_2 + 1) - T_1 = 9(18 + 1) - 94 = 77$$

The test statistic is

$$T = \min(T_1, T_1^*) = 77$$

With $n_1 = n_2 = 9$, the two-tailed rejection region with $\alpha = .05$ is found in Table 7b to be $T_1^* \leq 62$. The observed value, $T = 77$, does not fall in the rejection region and H_0 is not rejected. We cannot conclude that the distributions of brightness measurements differ for the two processes.

2 To test the null hypothesis that the two population means are identical, calculate

$$\bar{x}_1 = \frac{\sum x_{1j}}{n_1} = \frac{74.4}{9} = 8.2667 \qquad \bar{x}_2 = \frac{\sum x_{2j}}{n_2} = \frac{73.2}{9} = 8.1333$$

$$s^2 = \frac{(n_1 - 1)s_1^2 + (n_2 - 1)s_2^2}{n_1 + n_2 - 2} = \frac{625.06 - \frac{(74.4)^2}{9} + 599.22 - \frac{(73.2)^2}{9}}{16} = .8675$$

and the test statistic is

$$t = \frac{\bar{x}_1 - \bar{x}_2}{\sqrt{s^2\left(\frac{1}{n_1} + \frac{1}{n_2}\right)}} = \frac{8.27 - 8.13}{\sqrt{.8675\left(\frac{2}{9}\right)}} = .304$$

The rejection region with $\alpha = .05$ and 16 degrees of freedom is $|t| > 1.746$ and H_0 is not rejected. There is insufficient evidence to indicate a difference in the average brightness measurements for the two processes.

Notice that the nonparametric and parametric tests reach the same conclusions.

15.61 Since this is a paired experiment, you can choose either the sign test, the Wilcoxon signed rank test, or the parametric paired t test. Since the tenderizers have been scored on a scale of 1 to 10, the parametric test is not applicable. Start by using the easier of the two nonparametric tests – the sign test.

207

Define $p = P(\text{tenderizer A exceeds B for a given cut})$ and x = number of times that A exceeds B. The hypothesis to be tested is

$$H_0 : p = 1/2 \quad \text{versus} \quad H_a : p \neq 1/2$$

using the sign test with x as the test statistic. Notice that for this exercise $n = 8$ (there are two ties), and the observed value of the test statistic is $x = 2$.

***p*-value approach:** For the observed value $x = 2$, calculate

$$p\text{-value} = 2P(x \leq 2) = 2(.145) = .290$$

Since the p-value is greater than .10, H_0 is not rejected. There is insufficient evidence to indicate a difference between the two tenderizers.

If you use the Wilcoxon signed rank test, you will find $T^+ = 7$ and $T^- = 29$ which will not allow rejection of H_0 at the 5% level of significance. The results are the same.

15.63 To test for negative correlation with $\alpha = .05$, index .05 in Table 9, and the rejection region is $r_s \leq -.564$. The null hypothesis is rejected and we conclude that there is a negative correlation between the two variables.

15.65 The hypothesis to be tested is

H_0: population distributions 1 and 2 are identical
H_a: the distributions differ in location

and the test statistic is T, the rank sum of the positive (or negative) differences. The ranks are obtained by ordering the differences according to their absolute value. Define d_i to be the difference between a pair in populations 1 and 2 (i.e., $x_{1i} - x_{2i}$). The differences, along with their ranks (according to absolute magnitude), are shown in the following table.

d_i	−31	−31	−6	−11	−9	−7	7
Rank $|d_i|$	14.5	14.5	4.5	12.5	10.5	7	7

d_i	−11	7	−9	−2	−8	−1	−6	−3
Rank $|d_i|$	12.5	7	10.5	2	9	1	4.5	3

The rank sum for positive differences is $T^+ = 14$ and the rank sum for negative differences is $T^- = 106$ with $n = 15$. Consider the smaller rank sum and determine the appropriate lower portion of the two-tailed rejection region. Indexing $n = 15$ and $\alpha = .05$ in Table 8, the rejection region is $T \leq 25$ and H_0 is rejected. We conclude that there is a difference between math and art scores.

15.67 Similar to Exercise 15.39. The ranks within each block are shown below.

Varieties	1	2	3	4	5	6	T_i
A	4	2	3	3	3	3	18
B	1	3	2	2	2	2	12
C	5	4	4	4	4	5	26
D	3	5	5	5	5	4	27
E	2	1	1	1	1	1	7

a Use the Friedman F_r statistic calculated as

$$F_r = \frac{12}{30(6)}\left[18^2 + 12^2 + \cdots + 7^2\right] - 3(36) = 20.13$$

and the rejection region is $F_r > \chi^2_{.05} = 9.49$. Hence, H_0 is rejected and we conclude that there is a difference in the levels of yield for the five varieties of wheat.
b From Exercise 11.68, $F = 18.61$ and H_0 is rejected. The results are the same.

15.69 **a-b** Since the experiment is a completely randomized design, the Kruskal Wallis H test is used. The combined ranks are shown on the next page.

Plant	Ranks					T_i
A	9	12	5	1	7	34
B	11	15	4	19	14	63
C	3	13	2	9	6	33
D	20	17	9	16	18	80

The test statistic, based on the rank sums, is

$$H = \frac{12}{n(n+1)} \sum \frac{T_i^2}{n_i} - 3(n+1)$$

$$= \frac{12}{20(21)} \left[\frac{(34)^2}{5} + \frac{(63)^2}{5} + \frac{(33)^2}{5} + \frac{(80)^2}{5} \right] - 3(21) = 9.08$$

With $df = k - 1 = 3$, the observed value $H = 9.08$ is between $\chi_{.025}$ and $\chi_{.05}$ so that $.025 < p\text{-value} < .05$. The null hypothesis is rejected and we conclude that there is a difference among the four plants.

  c From Exercise 11.66, $F = 5.20$, and H_0 is rejected. The results are the same.

15.71   **a** Neither of the two plots follow the general patterns for normal populations with equal variances.

  b Use the Friedman F_r test for a randomized block design. The *Minitab* printout follows.

Friedman Test: Cadmium versus Harvest blocked by Rate

```
S = 10.33   DF = 2   P = 0.006

                        Sum
                         of
Harvest   N   Est Median   Ranks
1         6      202.29    11.0
2         6      201.21     7.0
3         6      300.73    18.0
Grand median = 234.74
```

Since the *p*-value is .006, the results are highly significant. There is evidence of a difference among the responses to the three rates of application.

15.73   The data are already in rank form. The "substantial experience" sample is designated as sample 1, and $n_1 = 5, n_2 = 7$. Calculate

$$T_1 = 19$$

$$T_1^* = n_1(n_1 + n_2 + 1) - T_1 = 5(13) - 19 = 46$$

The test statistic is

$$T = \min(T_1, T_1^*) = 19$$

With $n_1 = n_2 = 12$, the one-tailed rejection region with $\alpha = .05$ is found in Table 7a to be $T_1 \le 21$. The observed value, $T = 19$, falls in the rejection region and H_0 is rejected. There is sufficient evidence to indicate that the review board considers experience a prime factor in the selection of the best candidates.

15.75   Define $p = P(\text{student exhibits increased productivity after the installation})$ and x = number of students who exhibit increased productivity. The hypothesis to be tested is

        $H_0 : p = 1/2$ versus $H_a : p > 1/2$

using the sign test with x as the test statistic. Using the large sample approximation, the test statistic is

$$z = \frac{x - .5n}{.5\sqrt{n}} = \frac{21 - .5(35)}{.5\sqrt{35}} = 1.18$$

and the one-tailed rejection region with $\alpha = .05$ is $z > 1.645$. The null hypothesis is not rejected and we cannot conclude that the new lighting was effective in increasing student productivity.

15.77   Similar to Exercise 15.33. The data with corresponding ranks in parentheses are shown below.

Training Periods (hours)			
.5	1.0	1.5	2.0
8 (9.5)	9 (11.5)	4 (1.5)	4 (1.5)
14 (14)	7 (7)	6 (5)	7 (7)
9 (11.5)	5 (3.5)	7 (7)	5 (3.5)
12 (13)		8 (9.5)	
$T_1 = 48$	$T_2 = 22$	$T_3 = 23$	$T_4 = 12$
$n_1 = 4$	$n_2 = 3$	$n_3 = 4$	$n_4 = 3$

The test statistic, based on the rank sums, is

$$H = \frac{12}{n(n+1)} \sum \frac{T_i^2}{n_i} - 3(n+1)$$

$$= \frac{12}{14(15)} \left[\frac{(48)^2}{4} + \frac{(22)^2}{3} + \frac{(23)^2}{4} + \frac{(12)^2}{3} \right] - 3(15) = 7.4333$$

The rejection region with $\alpha = .01$ and $k - 1 = 3$ df is based on the chi-square distribution,

or $H > \chi^2_{.01} = 11.34$. Alternatively, the p-value can be bounded as $.05 < p\text{-value} < .10$. The null hypothesis is not rejected and we conclude that there is insufficient evidence to indicate a difference in the distribution of times for the four groups.

15.79   **a**   The quarterbacks have already been ranked. Let x be the true ranking and let y be my ranking. Since no ties exist in either the x or y rankings.

$$r_s = 1 - \frac{6 \sum d_i^2}{n(n^2 - 1)} = 1 - \frac{6 \left[(-2)^2 + (1)^2 + \cdots + (1)^2 \right]}{8(63)} = 1 - \frac{6(22)}{504} = .738$$

To test for positive correlation with $\alpha = .05$, index $.05$ in Table 9 and the rejection region is $r_s \geq .643$.

Alternatively, notice that the value of r_s for $\alpha = .025$ in Table 9 is exactly $.738$, so that p-value $= .025$. We reject the null hypothesis of no association at the 5% level, and conclude that a positive association exists.

CPSIA information can be obtained
at www.ICGtesting.com
Printed in the USA
FFOW05n0554110713
1374FF